SUSPENDED SOLIDS
IN WATER

MARINE SCIENCE

Coordinating Editor: Ronald J. Gibbs, *University of Delaware*

A Continuation Order Plan is available for this series. A continuation order will bring delivery of each new volume immediately upon publication. Volumes are billed only upon actual shipment. For further information please contact the publisher.

SUSPENDED SOLIDS IN WATER

Edited by

Ronald J. Gibbs

College of Marine Studies
University of Delaware
Lewes, Delaware

PLENUM PRESS • NEW YORK AND LONDON

Library of Congress Cataloging in Publication Data

Main entry under title:

Suspended solids in water.

(Marine science, v. 4)
"Proceedings of a symposium conducted by the Ocean Science and Technology Division of the Office of Naval Research . . . held in Santa Barbara, California, March 20-22, 1973."
Includes bibliographies.
1. Detritus — Congresses. 2. Marine sediments — Congresses. I. Gibbs, Ronald J., 1933-   ed. II. United States. Office of Naval Research. Ocean Science and Technology Division.
GC380.S97   551.4′601   74-19329
ISBN 0-306-35504-3

Proceedings of a symposium conducted by the Ocean Science and Technology
Division of the Office of Naval Research on Suspended Solids in Water
held in Santa Barbara, California, March 20-22, 1973

PREFACE

This book represents a compilation of papers presented at a symposium on the subject "Suspended Solids in Water," held in Santa Barbara, California, U.S.A., on March 20, 21 and 22, 1973. The symposium was sponsored by the Office of Naval Research and was designed to bring together a group that represented the dominant cross section of international research in this area. The idea for the conference originated as ONR recognized a potentially interesting area that, to date, had not had the benefit of a coordinating symposium and/or a book published on the specific subject.

In addition to the formal presentation of papers — informal open discussions followed — the symposium included two stimulating workshops. An abundance of impromptu exchange filled unscheduled periods. Many of the contributors have incorporated in their papers the ideas and points raised in discussions following formal presentation and at other times. The reader thus actually profits from the various discussions throughout the meeting.

The two half-day workshops were directed toward the subjects of sampling and concentrating suspended materials and *in situ* instrumentation. I have summarized the discussions from the two workshops for the reader and have correlated the material wherever possible in an introductory chapter. I have also included introductory material to acquaint the newcomer with the general field.

My sincere appreciation is extended to all who have contributed to the volume and to its preparation including the participants in the meeting, the reviewers of the papers and my wife, to whom was commissioned the preparation of the manuscripts for publication. Appreciation is also given to the Office of Naval Research and, in particular, to Alex Malahoff for his enthusiastic encouragement and support.

Ronald J. Gibbs

CONTENTS

I

PRINCIPLES OF STUDYING
SUSPENDED MATERIAL AND ITS SETTLING VELOCITIES

Principles of Studying Suspended Materials in Water

RONALD J. GIBBS

University of Delaware

ABSTRACT

The sampling methods for large-volume sampling both at surface and at depth are discussed. The advantages and disadvantages of centrifuging and filtering are covered, along with practical considerations of the various types of molecular filters. Various aspects of light transmission and scattering methods and their instrumentation are evaluated. The importance of and difficulty of adequate calibration is covered briefly.

INTRODUCTION

Information from the workshops on sampling and concentrating suspended material and on *in situ* instrumentation for the study of suspended material, from the literature, and from the author's experience is presented to review the basic approaches in the study of suspended materials in water. This presentation is not limited to the topics of two workshops of the symposium in the recognition that a fairly large portion of the readers of this volume will have broader interests than the specialized topics of the workshops. While the workshop discussions were directed mainly toward suspended material in the oceans, most of the information covered applies to the study of suspended material in most environments.

The scope of the problem and the variety of reasons that investigators want to study suspended solids in water should be considered. The approach to be used in studying suspended material should vary with the problem under study. What may be critical to one study may be insignificant to another. For example, depending

3

on the geographical and environmental locale, the concentration encountered can vary by up to 7 orders of magnitude, with a range from as much as 50,000 ppm observed in some rivers and estuaries to as little as 0.01 ppm in the open ocean depth. Likewise, suspended material varies in both size and composition from study to study. The open ocean surface and sewage outflow contain mainly organic material, whereas in rivers, estuaries, and at depth in the ocean, inorganic material is more important. Further, interest in size of material can vary from submicron material at depth in the open ocean to material having diameters of tens of microns in the surface ocean water (plankton) and to the fresh water and high energy marine areas where material may extend to a sand size of one millimeter. With this in mind, the immense bounds of the types of problems that can arise in the study of suspended material can be appreciated.

POINT-SAMPLING TECHNIQUES

Surface sampling techniques are very simple and usually involve either a container or pump. Containers for sampling range from simple buckets to some of the more complex containers used for sampling at depth. An important fact often overlooked in sampling the surface layer is that, because of surface tension, concentrations can vary widely. Because of this non-uniform distribution, great care in sampling is needed to obtain truly representative samples.

At depth, sampling techniques are more complicated. In the deep ocean, a large volume of sample is required. Successful point-sampling containers range in size from 1000-ℓ bags, to 200 to 300-ℓ tanks to 1 to 30-ℓ bottles. All these can be lowered by cable and are generally opened by sliding a messenger down the cable. Numerous commercial varieties of these samples are available.

Pumping systems are usually required for sampling material of low concentration. Large volumes — often of many hundred liters — must be processed in order to obtain samples of sufficient weight for analysis. The main criterion for selection of a pumping system is that it does not add significant contamination nor appreciably change the material in relationship to the aims of the project. Pumping systems have been used and appear to be practical for sampling from the surface to a few hundred meters depth. A variety of both submersible pumps and deck-type pumps have proven satisfactory. The danger of altering the size distribution is probably much higher with a pumping system than with any of the point-sampling devices mentioned. In using a pump system for measuring concentration and composition, the main concern, again, is whether the pump adds contaminants to the sample. The problem of contamination must be considered on an individual basis. In a study of trace metals, a few brass parts may add significant contamination, while the same pump system could be acceptable for a study of trace organic material.

In situ filtering is an interesting alternative to bringing large samples aboard a ship, particularly if the samples are obtained from great depth. Basically, these systems [*Spencer and Sachs*, 1970] consist of a battery in a special oil-filled container, a DC motor-pump, a flow meter, and a molecular filter. The intention is to filter a large volume — 200 to several thousand liters of water — *in situ*, obtaining a sample of the suspended material without handling this large volume of water. A second advantage is that this approach is much less prone to contamination. A photograph of the author's *in situ* filtering device is shown in Figure 1. A similar device was used by *Baker et al.*, as reported in this volume.

Fig. 1. *In situ* filtering device showing battery box on bottom; motor, pump, and discharge recording device above; and molecular filter holder at top.

CONCENTRATING THE SUSPENDED MATERIAL

Since the suspended material sample must be concentrated for weighing or for analysis, concentration of the material is a necessaryy step in all procedures. The two most popular concentration methods are centrifuging and filtering. The problems encountered here are vastly different, and depend on the concentration of the material in the samples.

The advantage of centrifuging when using the continuous-flow device is its easy use with large-volume samples. However, centrifuging has a distinct disadvantage when fine-grained material having organic material associated with it is being dealt with in that the density of some of the organic material can have a density close to that of water, making it very difficult to separate even at high speeds. The centrifuged samples would, therefore, be biased against this fine/light fraction and would give concentration values that are lower than in reality.

The use of filtering to extract suspended material from a sample is, by far, the most frequently used method. Numerous devices commercially available either suck the sample through the filter using a vacuum or force the sample through the filter using compressed gas. Two different types of molecular filter material are currently being used. Nuclepore-brand filters have the appearance of thin plastic with punched holes of specified diameters. Millipore and Gelman-brand filters, on the other hand, are much thicker and resemble an intricate cellular network. The pore sizes for the Millipore and Gelman-brand filters are specified as that size particle which will be trapped 100% by the filter. It should be remembered that 100% of the specified particle size will be retained on the filter, 80% of the next-smaller diameter particle, 50% of the next-smaller diameter particle, etc., will be retained on the filter.

Unless the concentration of suspended material is extremely low, both types of filters will give satisfactory results. Where low concentration occurs, as in the open ocean, the Nuclepore filter may be easier to use because of its lighter weight. This filter also allows less loss of weight due to wetting agents being removed by the water, has a smoother surface making work with a microscope easier, and gives lower chemical blanks for a number of elements. On the other hand, Millipore and Gelman filters give higher sustained flow rates and appear to give higher retention.

In using either type of filter in salt water, it should be remembered that, when dried, the salt water in the filter will add to the "sample weight". Therefore the salts must, obviously, be washed out of the filter before weighing. Five rinses using a small volume of distilled water that circulates to all parts of the filter are usually adequate. In determining concentration of suspended material by weighing the filter-plus-sample, where the weight of the filter is a reasonably large percentage of the total weight, great care must be exercised in dealing with the real weight of the filter. An incorrect filter weight can produce large errors for several reasons. Generally, filters are weighed prior to use and it is assumed that this weight does not change during use. However, since the filters have wetting agents and possible other materials that are affected by water when the filters are in use, they do lose weight resulting in erroneously large sample weights. Two methods can be used to overcome this problem successfully. The researcher can pre-wash filters before weighing and/or run a duplicate control filter beneath the regular filter to monitor the weight loss with use. Nuclepore-type filters appear to have a lower loss of weight than the Millipore and Gelman types.

IN SITU MEASUREMENT OF SUSPENDED MATERIAL

Numerous physical methods might be used to monitor suspended material. To date, however, only two methods have proven useful in measuring the *in situ* concentration of suspended material. These are measurement of the absorption of radioactive energy and the measurement of the adsorption and scattering of light.

Radioactive Absorption

The absorption of radiation is proportional to the mass present and therefore directly measures the concentration of suspended material. This method has been adapted using the radioactive isotopes Am-241 and Cd-109 as energy sources and placed in a fish-like probe for use in rivers and estuaries [*Florkowski and Cameron*, 1965]. However, the reader should be aware of several limitations. The system is not very sensitive and is, therefore, applicable only in the range of 0.5 to 10% concentration. However, this high (5000 ppm) threshold is encountered in only a few rivers and in some industrial uses. A further consideration is the effect of compositional changes of the suspended material on the absorption, which could be mistaken for changes in concentration. For example, if a suspension is suddenly changed from total quartz to total calcite and maintained 1000 ppm, the radioactive absorption would suggest that 2290 ppm were present. The method is satisfactory in studies with high concentrations of suspended material and where the mass absorption of the sediment to radiation (namely composition) does not vary with time.

Optical Absorption

Optical methods have been, by far, the most popular techniques used for studying suspended material as judged by the abundance of references on optical methods given in this volume. They have the sensitivity required to be useful in even the clearest open ocean water and, at the same time, are adaptable for use in high-concentration environments. However, the optical systems do have limitations, some of which are discussed by *Smith et al.* and *Tyler et al.* in this volume.

The absorption meter, frequently called a beam transmissometer or attenuance meter, has been used most often in the past to measure the concentration of suspended material mainly because of availability and simplicity. The transmissometer consists of a light source shining through a water path and projecting onto a sensor. The transmissometer has several limitations when used to determine the concentration of suspended material. Some dissolved substances do absorb light, especially the dissolved organic material so common to river water, coastal areas, sewerage, etc. Therefore, increased amounts of these organic dyes could be mistaken for suspended material. Several commercial companies produce transmissometers, suitable for field use, including
Hydro Products, P.O. Box 2528, San Diego, CA, 92112;

Martek, Inc., 877 W. 16th St., Newport Beach, CA. 92660;
Montedoro-Whitney, P.O. Box 1401, San Luis Obispo, CA. 93401;
As an example of the type of data that can be obtained using
the transmissometer and to point out the type of variation that can
be expected from changes in the size of suspended material, some actual working calibration curves are shown in Figure 2.

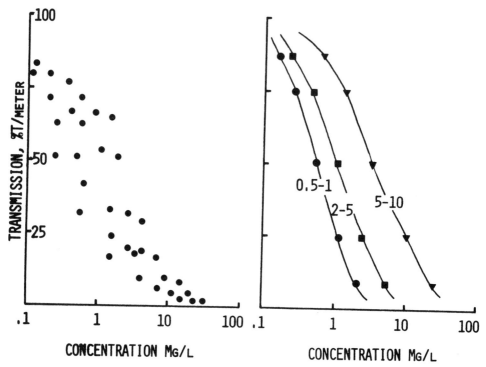

Fig. 2. Transmission calibration curve for absorption studies
of suspended solids in water. (*a*) Calibration curve for suspended
natural sediment obtained *in situ* from molecular filter sample at
the same place and time as the instrument reading; (*b*) Calibration
curves prepared with distilled water and a known amount of sized natural sediments.

The curve in Figure 2*b* shows some calibration curves for various size
fractions of sediments prepared in distilled water. These calibration curves are for material having approximately the same composition. It should be noted that changing composition can give a similar corresponding family of curves, as shown for changing particle
size. The graph in Figure 2*a* shows a curve derived from transmission readings and sampling in a field area. This field calibration
curve, which is fairly typical, represents material that has varying
size, composition, and shape as may occur in all non-controlled field
investigations. It should be noted that this is generally the type
of curve that will be obtained in any field study. In Figure 2*a* the

flattening out of the curve at about 80% transmission probably represents an absorption from dissolved organic material in the water
and can be considered a background below which accurate work cannot
be accomplished. The accuracy of a transmission curve is really
fairly poor. If a reading of 50% transmission is studied, it could
represent 0.2 to 2 mg/ℓ concentration, giving a 900% error. The
sensitivity of the transmissometer can be varied by changing the
length of the light path through the water. For example, in rivers
having 100 ppm suspended material, a light path of 25 cm is satisfactory. For the open ocean, a light path of 2 or 3 may be required.

Optical Scattering

In recent years, the use of light scattering has become the
meaningful method for measuring suspended material concentration.
In this method, the light that is scattered off suspended particles
at various angles to the oncoming light beam is measured. Dissolved
organic material that affects the transmissometer has no significant
effect on the scattering system. Further, the sensitivity of a
scattering system can be 0.01 ppm or lower. In this volume, the
details of the theory of light scattering are discussed by *Smith et
al.* [this volume] and an absolute light-scattering calibration technique is discussed by *Fry* [this volume]. Here, only the practical
limitations of light scattering will be given with comments on the
variety of *in situ* light-scattering meters in use.
The ideal case of light scattering is considering a suspension
of spheres. In a general sense, intensity is very high in the forward area at low angles from the direct beam. For a typical suspension (Figure 3) intensity will vary up to 6 orders of magnitude.
Likewise, it should be noted that the vast majority of scattered
light is in the region of low-angle forward scattering. If light
energy having a narrow band width is used and the sphere size is uniform, modes will also appear on this general distribution. Most
studies will not note these modes either because the light energy is
of a broader wave band or the particles do not all have the same size,
shape and index of refraction. The type of modes discussed is illustrated by a dotted line superimposed on the general curve in
Figure 3.
The factors that control the scattering of light are (1) concentration of suspended material, (2) size distribution of suspended material, (3) index of refraction of the suspended solid
relative to water, and (4) particle shape. While these four mechanisms can cause confusion if concentration alone is of interest,
they hold the possibility, if properly understood, of obtaining information about all four of the parameters.
The scattering response of a suspension to light varies with
suspended material of different particle sizes. The Mie theory
[*Mie*, 1908] can be used to calculate light scattered by spherical
particles of different sizes at different angles of scattering.
These data show that certain size ranges are responsible for the

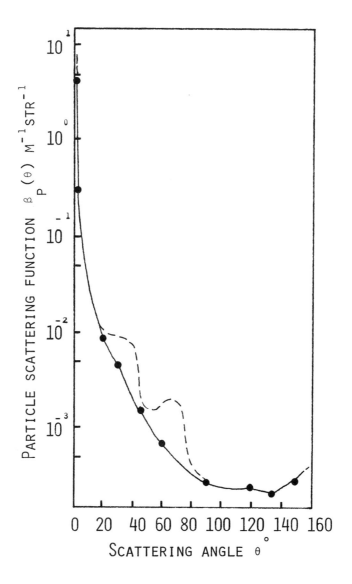

Fig. 3. Light-scattering intensity versus the scattering angle for a natural ocean sample. Dashed line shows typical Mie theory modes.

dominant light scattering at specific angles (Table 1).

Total Scattering

To measure the light scattered at all angles at the same time, an instrument that measures the integrated light from a low forward angle to the back scattering must be used. Several total-scatter-

TABLE 1.   The Ranges of Particle Size Contributing to the Middle
90% of the Total Scattering Coefficient and Volume-Scattering
Functions at Given Angles (After *Pak et al.* [1971]).

b	90% range	
	5%	95%
Total Scattering Coefficient	0.6μ	8.6μ
β 1°	1.3μ	18.5μ
β 5°	0.9μ	4.5μ
β 45°	0.3μ	8.5μ
β 90°	0.1μ	7.5μ

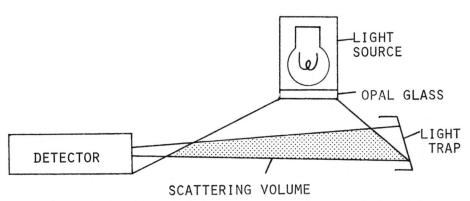

Fig. 4.   Drawing of a total scatterance meter after *Beutell and
Brewer* [1949].

ance meters have been used which are basically variations of a design by *Beutell and Brewer* [1949] shown in Figure 4.   It should be
noted that the total scatterance meter recieves almost all of its
signal from extremely low-angle forward scattering, seen in Figure
3.   In this volume *Baker et al.*, *Feely et al.*, *Kullenberg*, and
Smith et al. discuss or use total scatterance meters.

Specific-Angle Scattering

A wide variety of *in situ* scattering meters to measure light
scattering at various angles are in use.   The low-angle forward
scatterance meters have been the most popular specific-angle meter.
They range from the extremely low-angle (< 2°) forward scatterance
meters discussed by *Smith* in this volume and by *Kullenberg* [1968]
to the moderately low angle (5 to 23°) meter used by *Eittriem and
Ewing; Feely et al.;* and *Bassin* [all in this volume].   The basic
features of the design of the low-angle scatterance meter of
Kullenberg [1968] are shown in Figure 5a.   The moderately low-
angle scatterance meter shown in Figure 5b is similar to the design
of *Thorndike and Ewing* [1967] used by *Biscaye and Eittriem; Feely*

(*a*)

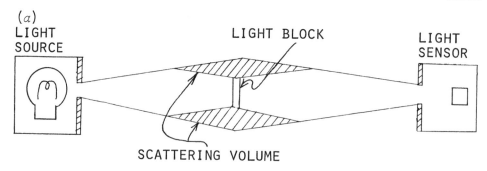

(*b*)

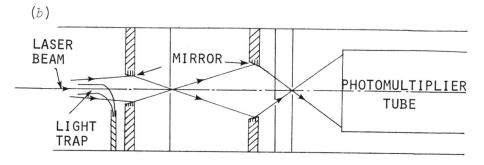

Fig. 5*a*. (Top) Drawing of an extremely low-angle scattering meter after *Kullenberg* [1968].

Fig. 5*b*. (Bottom) Drawing of a moderately low-angle scattering meter after the design of *Thorndike and Ewing* [1967].

et al.; and *Eittriem and Ewing*, all in this volume. The low-angle forward scattering meter takes advantage of the high intensity of scattered light relative to other angles of scattering and is more sensitive to the scatterance from larger particles.

Attempts to take advantage of the effects of particle size and composition on the scattering of light have lead several researchers to measure light scatterance at higher angles. The composition of the solid particles is inferred from the index of refraction between the water and the solid particles. Practically, this compositional differentiation is only plausible in distinguishing between organic material (refractive index: <1.1) and inorganic crystalline material (refractive index: >1.1). In this volume, *Gordon* and *Zaneveld* give two of the approaches used to determine the index of refraction of the solid suspended material.

An example of an instrument that measures light scattering at multiple angles is given in Figure 6. The author's instrument measures light scattering at adjustable low angles of forward scattering and at 45°, 90°, and 135° angles in addition to temperature, conductivity, and depth. This instrument has been tested to 15000 psi and can either record the data in the pressure case or can telemeter the data by cable.

Fig. 6. Multi-angle scattering meter developed by the author.

CALIBRATION OF INSTRUMENTS
FOR MEASURING SUSPENDED SOLIDS

The calibration of any instrument for measuring suspended solid material is critical to obtaining accurate information. The
purpose of each study dictates the type of calibration needed.
Since the optical data from a natural suspension is very complex
and is a function of the size, index of refraction and shape of
the solid particles, calibration can be a great source of error.

There are two basic approaches to the calibration. The suspended solids sampled or an adequate substitute can be used, or a
known light energy source or an adequate substitute can be used.
The latter method — using a known light evergy source — is generally used by researchers studying the optical properties of water and lends itself to intercalibration of instruments. The
difficulty of the first method — calibrating the instruments with
the actual material being studied — is that often the material is
difficult to obtain, either due to problems of sampling, concentrating, or analysis. It will be noted that the majority of authors
in this volume have prepared calibration curves by sampling while
obtaining instrument readings.

In studies of suspended solids, the only acceptable calibration method is calibration of results with the solids in suspension
that are actually under study. The use of material for calibration

other than that material being studied allows many sources of error. These errors — which can be as high as 1000% — are due to differences in size, composition, or particle shape between the actual suspended material and the material used in the calibration procedure. The reporting of data in Jackson Turbidity Units [*Whipple and Jackson*, 1900] is an example of using a physical property (the light from a special candle) that has little relationship to the suspended solids being calibrated. This outmoded method should be abandoned. A second method unacceptable for calibration in studies of suspended solids is the use of the polymer formazin [*American Public Health Association*, 1971] and the reporting of results in Formazin Turbidity Units. However, use of formazin for calibration can, again, allow errors of up to 1000% when related to the material in suspension. This error cannot be compensated for because it varies from study to study. The use of Formazin Turbidity Units in studies of suspended material can be equated to a chemist's reporting a millivolt reading instead of calibrating and reporting pH units.

CONCLUSION

With proper care and planning, suspended solids in water can be accurately studied either by direct sampling or by *in situ* instrumentation. Studies of suspended solid material have been successfully completed from high-concentration riverine and estuarine environments to the extremely low concentrations of the deep oceanic environment.

REFERENCES

American Public Health Association, *Standard Methods for the Examination of Water and Wastewater*, 13th ed., Washington, D.C., 874 pp., 1971.

Beutell, R.G., and A.W. Brewer, Instruments for the measurement of the visual range, *J. Sci. Inst.*, *26*, 357-359, 1949.

Florkowski, T., and J.F. Cameron, A simple radioisotope-ray transmission gauge for measuring suspended sediment concentrations in rivers, in *Proceedings of the Symposium on Radioisotope Instruments in Industry and Geophysics*, The International Atomic Energy Agency, Warsaw, 18-22 Oct. 1965, vol. 1, p. 395-410, 1965.

Kullenberg, G., Scattering of light by Sargasso Sea Water, *Deep Sea Res.*, *15*, 423, 1968.

Mie, G., Beitrage zur optik trüber Medien speziell Kolloidalen Metallosungen, *Ann. Physik.*, *25*, 377, 1908.

Pak, H., J.R.V. Zaneveld, and G.F. Beardsley, Jr., Mie scattering by suspended clay particles, *J. Geophys. Res.*, *76*, 5065-5069, 1971.

Spencer, D.W., and P.L. Sachs, Some aspects of the distribution, chemistry, and mineralogy of suspended matter in the Gulf of Maine, *Marine Geol.*, *9*, 117, 1970.

Thorndike, E.M. and M. Ewing, Photographic nephelometers for the deep sea, in *Deep Sea Photography*, edited by J.B. Hersey, Johns Hopkins University Press, Baltimore, Md., pp. 113-116, 1967.

Whipple, G.C. and D.D. Jackson, A comparative study of the methods used for the measurement of turbidity. *MIT Quart.*, *13*, 274, 1900.

Stokes' Settling and Chemical Reactivity of Suspended Particles in Natural Waters

ABRAHAM LERMAN, DEVENDRA LAL, AND
MICHAEL F. DACEY

*Northwestern University; Physical Research
Laboratory (India) and Scripps Institution
of Oceanography; Northwestern University*

ABSTRACT

*Equations are given for the Stokes settling velocities of the
following particle shapes: the sphere, oblate spheroid, prolate
spheroid, circular cylinder, elliptic cylinder, disc, and hemispherical cap. Dissolution of calcareous and silicate particles
settling through ocean water, based on literature data, is analyzed in terms of a model for dissolution rate independent of the
particle surface area, and a model for dissolution rate dependent
on a surface reaction. The settling of dissolving particles in the
presence of a countercurrent of upwelling water may lead to formation of thin nepheloid layers. Settling of calcite crystals
through a stratified water column is treated as a case of variable
nucleation (production) rates, dissolution and agglomeration of
crystals en route to the bottom. A stochastic model presented in
the paper gives a reasonably simple method for treating transient
transport of particles in a physically heterogeneous water column.*

INTRODUCTION

Chemical reactivity of the suspended matter in ocean water
manifests itself in uptake, release, and chemical exchange of
dissolved species, reactions between mineral phases and dissolved constituents, and decomposition and oxidation of organic
matter settling through the water column. A number of processes
that involve suspended matter in the ocean emphasize the importance
of biological/inorganic interactions in the formation and subsequent behavior of settling material. These are: the relatively
high abundance of particulate organic matter in ocean water
[*Manheim et al.*, 1970]; the formation of organic aggregates by
bacterial and inorganic processes [*Riley*, 1970]; the biological
productivity; uptake of trace metals by marine organisms [*Goldberg*,

1965, pp. 183-187; *Turekian et al.*, 1973]; oxidation of organic matter in the deeper ocean, resulting in the release of nutrient and trace elements [*Redfield et al.*, 1963; *Schutz and Turekian*, 1965; *Sorokin*, 1972]. The role of biological/inorganic interactions extends to the cycling and transport of chemical constituents between ocean water and ocean floor.

Inorganic particulate material transported to the ocean from the atmosphere, continents, and continental shelves [*Chester*, 1972; *Gibbs*, 1967; *Krey*, 1967; *Biscaye*, 1965] consists mostly of silicate minerals, some of which can react with the major ions in sea water and form new minerals.

In this paper we present some results concerning the behavior of particulate mineral matter of biogenic and inorganic origins in natural waters. Our approach, although by necessity limited by the data available for testing our models, is aimed at constructing meaningful quantitative models based on fundamental physical and chemical processes that are likely to involve suspended matter. A number of simplifications that had to be introduced into our models do not, hopefully, detract from the generality of our arguments and conclusions. This state-of-the-art account points out a number of avenues that can be pursued in the sampling of suspended matter in natural waters as well as in laboratory experiments.

FUNDAMENTALS OF SETTLING VELOCITIES

Settling velocity of submillimeter-size particles in water is commonly defined as a steady velocity that results from the balance between the forces of gravity F_G, buoyancy F_B, and resistance or drag F_D. The net gravitational force, $F_G - F_B$, is easy to compute for a rigid particle of known volume: it is the product of the particle volume, acceleration due to the force of gravity g, and the density difference between the particle and the medium. The resistance or drag force F_D, however, must be either computed or determined experimentally. Mathematical relationships for the drag forces and their values for a number of bodies of different shapes are available in the literature [*Allen*, 1968; *Taylor*, 1968]. *Hutchinson* [1967, pp. 245-305] has summarized information on the sinking rates of particles of some of the different shapes, as well as the experimental and observational data, through the early 1960's, on the sinking rates of planktonic organisms in lake waters.

Settling Velocities of Particles of Various Shapes

The relationships and explanatory notes are given for the settling velocities of several bodies, the shapes of which are similar to some of the particles encountered in natural waters: sphere, oblate spheroid, prolate spheroid, circular cylinder, elliptic cylinder, disc, and hemispherical cap.

Sphere. The settling velocity of the sphere is the most commonly used approximation to the settling velocity of mineral particles in water. After the terminal velocity has been reached, the balance of the net gravity and drag forces is (for example, *Weber* [1972, p. 113])

$$F_G - F_B = F_D,$$

or

$$4\pi r^3 g(\rho_s - \rho)/3 = 6\pi\eta r U, \tag{1}$$

where r is the radius of the sphere, g is acceleration due to gravity, ρ_s is the particle density, ρ is the density of the medium, and η is viscosity. The settling velocity U, from (1), is

$$U = \frac{2g(\rho_s - \rho)r^2}{9\eta} = Br^2 \tag{2}$$

This relationship for U is Stokes' law for settling spheres, where we denote

$$B = \frac{2g(\rho_s - \rho)}{9\eta} . \tag{3}$$

For particles of density $\rho_s \simeq 1.5$ $g \cdot cm^{-3}$ in sea water, the value of B is

$$B = \frac{2 \times 981 \times 0.5}{9 \times 0.015} = 7.27 \times 10^3 \ cm^{-1} sec^{-1}.$$

The value of the coefficient B primarily depends on water temperature (viscosity increases by a factor of 1.6 in the temperature range from 5° to 25°C, see Figure 7b) and on the density of settling material (the difference $\rho_s - 1.0$ is strongly affected by the ρ_s values in the range from 1.X to 2.X).

Oblate Spheroid and Prolate Spheroid. In an oblate spheroid the polar semi-axis c is shorter than the equatorial radius a, whereas in a prolate spheroid c is longer than a. The volume of either spheroid is $4\pi a^2 c/3$. The drag force of a spheroid oriented with its c-axis parallel to the flow direction U is [*Brenner*, 1964]

$$F_D = 6\pi\eta a U(1 - 0.2E) , \tag{4}$$

where $E = 1 - c/a$ is a measure of distortion of the sphere. Other relationships for the drag force have been given by *Payne and Pell* [1960] and *Breach* [1961]. In a sphere, $E = 0$. In an oblate spheroid, $1 > E > 0$. In a prolate spheroid, $E < 0$. In (4), the terms of second order in E have been omitted. The settling rate of a spheroid oriented with its polar axis parallel to the direction of settling is

$$U = \frac{2g(\rho_s - \rho) (1 - E)a^2}{9\eta(1 - 0.2E)} . \tag{5}$$

In considering the settling velocities of bodies of different shape, it is convenient to refer their volumes to an equal volume of a sphere of radius r. For a spheroid (a, c, $E = 1 - c/a$) equal in volume to a sphere r,

$$a = r(1 - E)^{-1/3} , \qquad (6)$$

and the settling velocity of such a spheroid is

$$U = \frac{2g(\rho_s - \rho)(1-E)^{1/3} r^2}{9\eta(1 - 0.2E)} . \qquad (7)$$

In a spheroid oriented with its polar axis c perpendicular to the direction of settling, the drag force is [Brenner, 1964]

$$F_D = 6\pi\eta a(1 - 0.4E) , \qquad (8)$$

and the settling velocity is

$$U = \frac{2g(\rho_s - \rho)(1 - E)a^2}{9\eta(1 - 0.4E)} . \qquad (9)$$

The settling velocity U, written in terms of the radius r of a sphere of equal volume is

$$U = \frac{2g(\rho_s - \rho)(1 - E)^{1/3} r^2}{9\eta(1 - 0.4E)} . \qquad (10)$$

It should be noted that as E tends to 0, the relationships for the drag force and settling velocity of the spheroids tend to the Stokes law relationships for a sphere, (1) and (2).

Hemispherical Cap. *Payne and Pell* [1960] have derived a relationship for the drag force on a hemispherical cap (calotte) oriented with the plane of its equatorial circle facing the flow:

$$F_D = 5.579\pi\eta aU , \qquad (11)$$

where a is the hemisphere radius.

To compare the hemispherical cap with a particle of similar shape and finite mass, we assume that the cap wall thickness is a small fraction of its radius. The volume of the wall material in such a cap, of outer surface radius a and inner surface radius a_i, is

$$V = 2\pi(a^3 - a_i{}^3)/3 = 2\pi a^3 q_r/3 , \qquad (12)$$

where $q_r = 1 - (a_i/a)^3$ was introduced for brevity, and $q_r < 1$.

Using (11) and (12), the settling velocity of the hemispherical cap is

$$U = \frac{2g(\rho_s - \rho)q_r a^2}{16.737\eta} \quad . \tag{13}$$

In terms of a sphere r volume equal to the cap wall volume,

$$a = r(2/q_r)^{1/3} \quad , \tag{14}$$

and the settling velocity is

$$U = \frac{2g(\rho_s - \rho)q_r^{1/3}r^2}{10.544\eta} \quad . \tag{15}$$

Disc. For a disc of radius a and "no thickness" [*Lamb*, 1932, p. 604; *Payne and Pell*, 1960; *Brenner*, 1964], the drag force of the disc facing the stream broadside is

$$F_D = 5.1\pi\eta a U. \tag{16}$$

When the disc faces the stream edgewise, the drag force is

$$F_D = 3.396\pi\eta a U. \tag{17}$$

Assuming, as an approximation, that the drag forces do not change for a slowly settling thin disc of thickness h ($q_h = h/a$ << 1; disc volume = $\pi a^2 h$), the settling velocities for the two orientations are

$$U = \frac{g(\rho_s - \rho)q_h a^2}{5.1\eta} \quad \text{(broadside)}, \tag{18}$$

$$U = \frac{g(\rho_s - \rho)q_h a^2}{3.396\eta} \quad \text{(edgewise)} . \tag{19}$$

In terms of a sphere volume equal to the disc volume, the radii a and r are related through

$$a = r\left[\frac{4}{3q_h}\right]^{1/3} \quad , \tag{20}$$

and the settling velocities become

$$U = \frac{g(\rho_s - \rho)q_h^{1/3}r^2}{4.21\eta} \quad \text{(broadside)}, \tag{21}$$

$$U = \frac{g(\rho_s - \rho)q_h^{1/3}r^2}{2.803\eta} \quad \text{(edgewise)} . \tag{22}$$

Circular Cylinder. The drag force per unit length on a circular cylinder (of radius a) immersed in a flow characterized by low values of the Reynolds number ($Re \ll 1$; where $Re = 2a\rho U/\eta$) is given by *Lamb* [1932, p. 616] and by *Munk and Riley* [1952, pp. 238-239]

$$\frac{4\pi\eta U}{2.00 - \ln Re} .$$

We assume, as an approximation, that the drag force relationship given above applies to a cylinder of some finite characteristic length L. Then, the drag force can be written as

$$F_D = \frac{4\pi\eta UL}{2.00 - \ln Re} . \tag{23}$$

Taking $L = 2a$, and the cylinder volume $\pi a^2 h$, the settling velocity becomes

$$U = \frac{g(\rho_s - \rho)(2.00 - \ln Re)ah}{8\eta} . \tag{24}$$

The cylinder radius a, written in terms of the radius r of a sphere of equal volume, is

$$a = r \left(\frac{4}{3q_h}\right)^{1/3} , \tag{25}$$

where $q_h = h/a$ is the height-to-radius ratio. The settling velocity is

$$U = \frac{g(\rho_s - \rho)(2.00 - \ln Re)q_h^{1/3}r^2}{6.604\eta} . \tag{26}$$

It should be noted that while (25) for a circular cylinder is identical to (20) for a thin disc, the relationships for the settling velocities of the two bodies differ.

Elliptic cylinder. The elliptic cylinder, having semi-axes a and b and height h, occupies the volume πabh. The drag force per unit length of such a cylinder, oriented with its a-axis parallel to the stream direction has been given by *Lamb* [1932, p. 617]. As in the case of the circular cylinder, we assume a characteristic length L for the cylinder. Then, the drag force is

$$F_D = \frac{4\pi\eta UL}{a/(a+b) - 0.577 + \ln[16/(1 + b/a)] - \ln Re} , \tag{27}$$

where the Reynolds number Re was defined in the preceding section on circular cylinder. Using $L = 2b$ in (27), the settling velocity is

$$U = \frac{g(\rho_s - \rho)\left[\frac{a}{a+b} - 0.577 + \ln\frac{16}{1+b/a} - \ln Re\right]ah}{8\eta} \quad \cdot(28)$$

It can be verified by inspection of the terms that, for an elliptic cylinder characterized by $b < a$, the drag force is smaller and the settling velocity is therefore greater than for a circular cylinder $(b=a)$. The curves in Figure 1 bear this out.

To write a relationship for the settling velocity of an elliptic cylinder in terms of the radius of a sphere of equal volume, we let $q_b = b/a$ and $q_h = h/a$. Then,

$$a = r\left(\frac{4}{3q_b q_h}\right)^{1/3}, \tag{29}$$

and the settling velocity becomes

$$U = \frac{g(\rho_s - \rho)\left[2.196 + \frac{1}{1+q_b} - \ln(1+q_b) - \ln Re\right]q_b^{-2/3}q_h^{1/3}r^2}{6.604\eta}, \tag{30}$$

Comparison of the settling rates. The settling velocities of the particles of different shapes but equal in volume, as given above and summarized in Table 1, have been computed for the range of equivalent sphere radii from 0.3 μm to 30 μm and plotted as shown in Figure 1. These sizes are within the range of the Stokes law obeyed by settling spheres [*Gibbs et al.*, 1971, with references to earlier literature]. The settling velocities of all the particle shapes considered, with the exception of the cylinders, vary linearly with r^2. For cylinders, the proportionality relationship approaches $r^{1.6}$. In the log-log coordinates of Figure 1, the log of the settling velocity increases linearly with log r. For the cylinders, the slight deviation from linearity in the log U against log r plot is due to the terms $\ln Re$ in (26) and (30).

It is interesting to note that the settling velocities of the spheroids characterized by pronounced oblateness $(c/a = 0.5)$ or prolateness $(c/a = 1.5)$ differ only little from the settling velocity of the sphere $(c/a = 1)$ of the same volume. The settling velocities in Figure 1 support the case of polar axes parallel to the settling direction $(U \parallel c)$. For each of the spheroids, the difference in the settling velocity owing to a different orientation (i.e., polar axes either parallel or perpendicular to the settling direction) are small and almost equal to the differences between their settling velocities shown in Figure 1 and the settling velocity of the sphere.

The settling velocities of the disc in two orientations and of the hemispherical cap are considerably smaller than the settling velocity of the sphere. The settling velocities of the cylinders are higher than the settling velocity of the sphere. Data presented

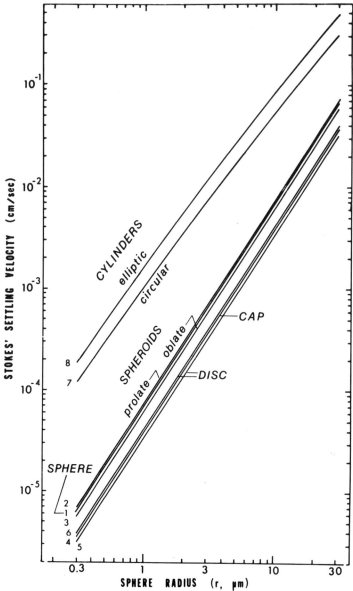

Fig. 1. Stokes settling velocities of particles of different shapes. The horizontal axis is the radius of the sphere equal in volume to the particle of different shape. The term $g(\rho_s - \rho)/\eta = 3.27 \times 10^4$ (where $\rho_s - \rho = 0.5$ and $\eta = 0.015$) for all the curves. 1. Sphere; 2. Prolate spheroid, $c/a = 1.5$, settling parallel to polar axis c; 3. Oblate spheroid, $c/a = 0.5$, same settling as in 2; 4. Hemispherical cap, wall thickness is 10% of the outer radius ($a_i/a = 0.9$, $q_r = 0.271$); 5. Disc, thickness to radius ratio $q_h = 0.1$ settling broadside; 6. Disc, same as in 5, settling edgewise; 7. Circular cylinder, height-to-radius ratio $q_h = 4$; 8. Elliptic cylinder, $q_b = 0.5$, $q_h = 4$, settling parallel to a-axis.

TABLE 1.   Stokes Settling Velocities of Some Axisymmetrical
Bodies U in Terms of the Radius r of the Sphere Equal
in Volume to the Other Body.

Particle shape	Characteristic dimension, in terms of equal volume sphere	$\dfrac{U\eta}{g(\rho_s - \rho)}$ $[cm^2]$
Sphere	r	$0.222r^2$
Oblate spheroid, polar semiaxis c, equatorial radius a. $c < a$, $E = 1 - c/a$. Prolate spheroid, $c > a$.	$a = r(1 - E)^{-1/3}$	$\dfrac{0.222(1-E)^{1/3}r^2}{(1 - 0.2E)}$ $(U \parallel c)$ $\dfrac{0.222(1-E)^{1/3}r^2}{(1 - 0.4E)}$ $(U \perp c)$
Disc, radius a. Approximation: thickness h. $q_h = h/a \ll 1$.	$a = r\left(\dfrac{4}{3q_h}\right)^{1/3}$	$0.238q_h^{1/3}r^2$ (broadside) $0.357q_h^{1/3}r^2$ (edgewise)
Hemispherical cap, radius a. Approximation: outer wall radius a, inner wall radius a_i. $q_r = 1 - (a_i/a)^3$	$a = r(2/q_r)^{1/3}$	$0.190q_r^{1/3}r^2$
Circular cylinder, radius a, height h, $q_h = h/a$.	$a = r\left(\dfrac{4}{3q_h}\right)^{1/3}$	$0.151(2.00 - \ln Re)q_h^{1/3}r^2$
Elliptic cylinder, semiaxes a and b, height h, $b \gtrless a$. $q_b = b/a$, $q_h = h/a$.	$a = r\left(\dfrac{4}{3q_b q_h}\right)^{1/3}$	$0.151[2.196 + \dfrac{1}{1+q_b} - \ln(1+q_b)$ $- \ln Re] \times q_b^{-2/3}q_h^{1/3}r^2$ $(U \parallel a)$

by *Weber* [1972, p. 114] indicate that, in a low Reynolds-number flow ($Re \ll 1$) the settling velocity of the cylinder is greater than that of the sphere.

Applicability to Natural Conditions

Reynolds number. The Stokes settling velocity of particles is a valid model for a low Reynolds number-flow ($Re \ll 1$). Using the definitions of the Reynolds number given in the preceding section, $Re = 2r\rho U/\eta$, and Stokes' settling velocity, $U = Br^2$, one can write $Re = 2\rho Br^3/\eta$.

$$Re = 2\rho Br^3/\eta = 2 \times 1 \times 7 \times 10^3 \times r^3/0.015 \simeq 10^6 r^3 \quad,$$

where r is in cm. Thus, even for particles of radius $r = 30\mu m$, the Reynolds number is $Re < 0.05$, and it assumes lower values for smaller particles.

Eddy turbulence. Retardation of the settling velocities of rain drops and solid particles by atmospheric turbulence (for example, *Smith* [1959, p. 196]) suggests that analogies in natural waters should be recognized. Some earlier works have pointed out the relationships between the sizes of turbulent eddies and the sizes of settling particles that may be affected by them [*Munk and Riley*, 1952, p. 233; *Levich*, 1962, p. 216]. Here, we present two arguments based on simplified relationships between the Stokes settling velocities and "macroscopic" eddy diffusivity.

Considering distances in the ocean water column of the order of 1 km, a particle of radius $r = 1$ μm settles through 1 km in time $t = 44$ years. If the particle transport were controlled only by eddy diffusion, an estimate of dispersal distance over the period of time t may be taken as $(Kt)^{1/2}$, where K is the eddy diffusion coefficient. For $t = 44$ yr and $K \simeq 1$ cm^2·sec^{-1} [*Munk*, 1966; *Craig*, 1969; *Lerman*, 1971], the distance $(Kt)^{1/2} = 0.37$ km is shorter than the distance covered by the Stokes settling. Thus, for particles of radius 1 μm and larger, the Stokes settling velocity dominates over eddy diffusional transport.

Another line of argument may be based on the settling flux. Considering the flux of settling particles as comprising an eddy diffusional and the Stokes settling contributions (for example, *Pasquill* [1962, p. 112]),

$$\text{Flux} = -K \frac{\Delta N}{\Delta z} + UN,$$

where N is the particle concentration, the relative magnitudes of the Stokes settling velocity $U = Br^2$ and the term $K/\Delta z$ can be compared. Assuming that the concentration gradient is linear over the distance Δz, and the values of ΔN and N are of the same order of magnitude, the terms $K/\Delta z$ and U represent two different velocities in the flux equation. Using, as before, $\Delta z = 1$ km, $K = 1$ cm^{2}sec^{-1}, and $U = 7.3 \times 10^3 r^2$ cm.sec^{-1}, the values $K/\Delta z = 1 \times 10^{-5}$ cm.sec^{-1} and $U = 7.3 \times 10^{-5}$ cm.sec^{-1} are of the same order of mag-

nitude. Thus, particles larger than 0.5-1 μm in radius are controlled in their settling primarily by the Stokes settling velocity.

In the remainder of this paper, discussion of the behavior of
settling particles will deal only with the Stokes settling velocity
and its contribution to the particle fluxes.

SETTLING OF DISSOLVING PARTICLES

Observations of man-produced ^{14}C activity in biogenic calcareous particles occurring in sea water down to 3500 m depth [Soma-
yajulu et al., 1969; Williams et al., 1970] are direct indication
of transport from the ocean surface to the greater depths by means
of settling material. The ^{14}C-ages of the biogenic material sampled at 2300 m and 3500 m depth indicate that the settling velocities of particles diminish at greater depths due to dissolution
[Somayajulu et al., 1969]. Additional evidence of dissolution has
been provided by the experiments of Peterson [1966] on dissolution
of calcite spheres suspended in the ocean at different depths, and
by similar experiments with calcareous skeletons of foraminifera
and siliceous skeletons of radiolarians [Berger, 1967, 1968].

Mechanisms and Rates of Dissolution

Experimental results of several investigators [Garrels and
Dryer, 1952; Nielsen, 1964, pp. 72-85, 120-126; Plummer, 1972] indicate that the rates of dissolution (and precipitation) of a number
of common mineral phases are power functions of the distance from
saturation. That is, the rates of dissolution are proportional to
$(C_{sat} - C)^n$, where C_{sat} is a saturation concentration, C is concentration in solution varying with time, and n is a positive number.

The case of $n = 1$ represents a first-order dissolution rate,
interpreted as a diffusion-controlled dissolution process [Berner,
1971, p. 65]. First-order dissolution kinetics has been shown to
apply to the rates of removal of Ca^{2+} from calcite, and removal of
SiO_2 from several clay minerals in laboratory-controlled dissolution experiments [Lerman et al., 1973].

The data of Peterson [1966] and Berger [1967] on in situ dissolution of calcite spheres and foraminifera indicate nearly constant rates of dissolution down to about 3500 m depth. Below 3500 m,
the rates of dissolution increase with depth, by as much as eight-
fold from 3500 m to 4500 m depth. The depth at which the dissolution rate begins to increase has been related to departures in the
pH from the calcite saturation values [Morse and Berner, 1972].

For sea water above 3500 m, the rate of dissolution of particles expressed as a loss of mass (in units of $g.cm.^{-2}yr^{-1}$) is nearly constant. Assuming that dissolving particles are spherical, the
rate of dissolution can be written as

$$dr/dt = -\varepsilon'/\rho_s \equiv -\varepsilon , \qquad (31)$$

Where r is the particle radius, ε' is dissolution rate in $g.cm.^{-2}$ yr^{-1}, ρ_s is the particle density, and ε is dissolution rate in units of $cm.yr^{-1}$ [Lal and Lerman, 1973].

The following values of the dissolution rate (ε) have been computed from the literature data: $0.15\times10^{-4}cm.yr^{-1}$ (data of Peterson [1966] on calcite spheres); 3.4×10^{-4} $cm.yr^{-1}$ (data of Berger [1967], on foraminifera); and $7\times10^{-4} - 10\times10^{-4}$ $cm.yr^{-1}$ (data of Morse and Berner [1972] on calcite powder in sea water). These computed rates of dissolution apply to spherical particles; for other shapes, it is conceivable that particle surface area and shape change non-uniformly as dissolution proceeds.

For siliceous particles, the rate of dissolution $\varepsilon = 1\times10^{-3} - 2.5\times10^{-3}cm.yr^{-1}$ has been computed from Berger's [1968] data on radiolarian skeletons. Dissolution rates 100 - 1000 times higher have been reported by Hurd [1972] from laboratory experiments with siliceous shells from deep ocean sediments.

A different mechanism of dissolution of a silicate phase has been proposed by Wollast [1967] in a study of dissolution of potassium feldspar ($KAlSi_3O_8$). In contact with water, $KAlSi_3O_8$ is transformed into another phase, poorer in SiO_2, that grows from the grain surface *inward*. In the process of this reaction, silica diffuses from the fresh feldspar surface through the altered layer and into the external solution. Thus, the dissolution-alteration process described by Wollast [1967] can be visualized as nearly conserving the original dimensions of a feldspar grain, yet contributing silica to solution in contact with the grain. Derivation of a relationship between the radius of a fresh mineral core in the grain and the time of contact (or settling distance) in a water column follows.

The mass of silica ($-dM$) removed from the fresh mineral surface, assumed spherical, is

$$-dM = -4\pi r^2 C_f dr , \qquad (32)$$

where C_f is the concentration of SiO_2 in the fresh mineral.

When transport of silica outward through the altered layer takes place by molecular diffusion, the amount removed ($-dM$) within time interval dt is the silica flux from the surface area of the fresh core,

$$-dM = \frac{D(C_s - C_w)}{r_o - r} \times 4\pi r^2 dt , \qquad (33)$$

where D is the diffusion coefficient of SiO_2 through the altered layer, $r_o - r$ is the thickness of the altered layer (r_o is the particle radius and r is the radius of the fresh mineral core), C_s is the concentration of SiO_2 at the fresh surface, and C_w is the SiO_2 concentration in water in contact with the mineral particle. The term

$D(C_s - C_w) / (r_o - r)$ in (33) is the diffusional flux due to a linear concentration gradient.

Division of (33) by (32) gives

$$\frac{dr}{dt} = - \frac{D(C_s - C_w)}{C_f(r_o - r)} . \tag{34}$$

Integration of (34) gives an explicit relationship between the diminishing radius of the fresh mineral core and the time or distance of settling. Since the settling velocity of a particle in the z direction is $dz/dt = U = Br_o^2$ as given in (2), where r_o is the particle radius, integration of (34) with respect to dz gives

$$r = r_o - (Az)^{\frac{1}{2}}/r_o , \tag{35}$$

where $A = 2D(C_s - C_w) / (C_f B)$; density is unaffected by dissolution.

The rate of decrease in the radius of fresh mineral core, $-dr/dt$ in (34), is not constant but it decreases with depth as the core radius r becomes smaller. A mean rate of dissolution $\bar{\varepsilon}$ of the fresh mineral core in a particle settling from depth z_1 to z_2 is

$$\bar{\varepsilon} = \frac{1}{z_2 - z_1} \int_{z_1}^{z_2} \left(- \frac{dr}{dt} \right) dz . \tag{36}$$

Using (35) in (34) and (36), the integration gives

$$\bar{\varepsilon} = \frac{A^{\frac{1}{2}} Br_o}{z_2^{\frac{1}{2}} + z_1^{\frac{1}{2}}} . \tag{37}$$

For potassium feldspar, the mean rate of dissolution between the depths $z_1 = 1$ km and $z_2 = 4$ km was computed using the following data: diffusion coefficient of SiO_2 through the altered surface layer $D = 1 \times 10^{-14}$ cm.^{2}sec^{-1}, and $C_s = 70$ mg SiO_2/ ℓ [Wollast, 1967]; $C_f = 1.62$ g SiO_2 per 1 cm^3 of feldspar; feldspar density $\rho_s = 2.5$ g.cm^{-3}; silica concentration in sea water $C_w = 6$ mg SiO_2/ℓ; and the particle radius $r_o = 4\mu$m. Using these values,

$$\bar{\varepsilon} = \frac{(1.09 \times 10^{-22})^{\frac{1}{2}} \times 7.27 \times 10^3 \times 3.16 \times 10^7 \times 4 \times 10^{-4}}{(4 \times 10^5)^{\frac{1}{2}} + (1 \times 10^5)^{\frac{1}{2}}} = 1 \times 10^{-6} \text{cm.yr}^{-1}.$$

The rate of dissolution of feldspar grains derived above is so much lower than the rates of dissolution of biogenic siliceous particles, that the contribution of silica to ocean water due to alteration of feldspar in the course of settling is probably negligible compared with the biogenic sources. Potassium feldspar, a com-

mon constituent of crustal rocks, has been reported as a common
constituent of the 2-20μm-size fraction of sediments in the Atlantic [*Biscaye*, 1965] and in suspended matter in the Caribbean
[*Jacobs and Ewing*, 1965].

Effect of Dissolution on Settling Velocities

If the size of a settling particle decreases owing to dissolution, its settling velocity U also decreases. It has been shown
[*Lal and Lerman*, 1973] that for spherical particles dissolving at a
constant rate of ε cm.yr^{-1}, the rate of decrease in the particle
radius r with depth z is $dr/dz = -\varepsilon/U$, and the relationship between
the radius and settling depth is

$$r = r_O \left[1 - \frac{3\varepsilon(z - z_O)}{Br_O^3} \right]^{1/3} , \qquad (38)$$

where r_O is the initial radius of the particle at depth z_O, B is the
Stokes settling parameter for a sphere defined in (3), and depth z
is positive and increasing downward from z_O.

If particles settle and dissolve through an upwelling water
mass, their settling velocity is the difference between the Stokes
settling velocity and the upwelling velocity of water, $U - U_\alpha$. Continuous dissolution of particles makes their Stokes settling velocity U equal to the advective velocity of rising water U_α at some
depth. After reaching that depth, the dissolving particles would
rise ($U < U_\alpha$) until their dissolution is complete at some depth
above the depth of reversal in the direction of settling. In such
a system, the particle radius as a function of depth is given by
[*Lal and Lerman*, 1973]

$$r^3 - \frac{3U_\alpha r}{B} - r_O^3 + \frac{3U_\alpha r_O + 3\varepsilon(z - z_O)}{B} = 0 . \qquad (39)$$

A number of conclusions arise from the equality of the settling
and upwelling velocities, $Br^2 = U_\alpha$, and from (39).

First, the radius of the particle at the depth of reversal r_r
is

$$r_r = (U_\alpha/B)^{\frac{1}{2}} , \qquad (40)$$

which shows that it is independent of the initial radius and independent of the rate of dissolution.

Second, the depth z_r where the reversal of the settling direction takes place depends on the rate of dissolution but, even more
strongly, on the initial radius of the particle:

$$z_r = \frac{2B}{3\varepsilon}\left(\frac{U_a}{B}\right)^{3/2} + \frac{Br_o^3}{3\varepsilon} - \frac{U_a r_o}{\varepsilon} . \tag{41}$$

Third, the depth where dissolution is complete ($z_{r=0}$) is shallower than the depth of reversal z_r, as can be verified by setting $r = 0$ in (39). The thickness of the water layer within which the particles reverse their direction of settling and dissolve completely is independent of the initial particle radius r_o, as obtained from (39) and (41),

$$z_r - z_r = 0 = \frac{2U_a(U_a/B)^{\frac{1}{2}}}{3\varepsilon} . \tag{42}$$

Equations (40) - (42) indicate that for larger particles the depths of reversal are greater, but the particle radius at the depth of reversal r_r and the thickness of the reversal layer ($z_r - z_{r=0}$) are the same as long as the physical conditions in the water column, reflected in the values of U_a and B, remain unchanged. These conclusions are shown graphically in Figures 2a and 3, and in the following computational example.

To compute the values of r_r and $z_r - z_{r=0}$, we take $U_a = 1.4\times10^{-5}$ cm. sec $^{-1}$ [cf. *Munk*, 1966; *Craig*, 1969], $B = 7.27\times10^3$ cm $^{-1}$ sec $^{-1}$, and $\varepsilon = 1.5\times10^{-5}$ cm.yr $^{-1}$ in (40) and (42), obtaining

$$r_r = 0.44\mu m, \quad \text{and} \quad z_r - z_{r=0} = 8.6 \text{ m}$$

Figure 3 shows the difference between the settling paths of dissolving particles in the presence and the absence of upwelling velocity of water (U_a). When $U_a = 0$, the particle dissolves within 22 m below the depth $z_{r=0}$, whereas, in the presence of upwelling, it is confined to about 9 m below $z_{r=0}$. In Figure 2a, the reversal layers are of the same thickness (8.6 m, on a log scale), but they occur at different depths owing to the fact that bigger particles settle farther before their size decreases to $r_r = 0.44\mu m$.

Particle Concentration and Dissolution

In a steady-state case, when the supply of particles at the upper boundary of the water column is constant, retardation of their settling rates owing to dissolution results in an increase in the concentration of particles (N, particles/cm^3) with depth:

$$N = N_o r_o^2/r^2 , \quad \text{or} \quad N = N_o(Br_o^2 - U_a)/(Br^2 - U_a) , \tag{43}$$

where the subscript o denotes the values of the parameters at the upper boundary of the water column z_o.

Relationships for the concentration of settling matter (S, mass/vol.), concentration of particles (N, particles/vol.), and rate of

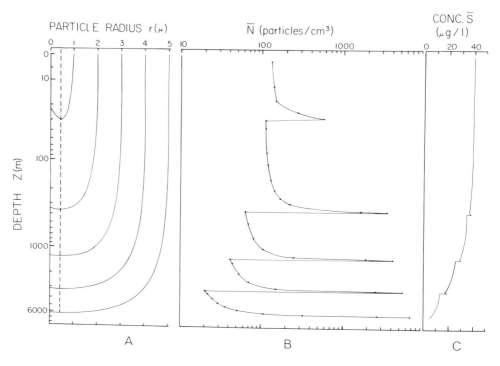

Fig. 2. Particle radii and concentrations, as a function of depth, in a heterogeneous population of dissolving particles settling through an upwelling water mass. At $z = 0$, the following steady concentrations, in units of particles/cm^3, and radii, in microns, are maintained: 25(1), 50(2), 25(3), 19(4), 13(5). Such a sample of spherical particles, of density 2.5g/cm^3, gives the concentration of suspended matter $S_0 = 40$ µg/ℓ. For all curves shown: dissolution rate $\varepsilon = 1.5 \times 10^{-5}$ cm.yr^{-1}, upwelling velocity of water $U_\alpha = 1.4 \times 10^{-5}$ cm.sec^{-1} ($= 4.42$ m.yr^{-1}), and the Stokes settling parameter $B = 7.27 \times 10^3$ cm^{-1}sec^{-1}.

(a) Decrease in particle radius, as a function of depth, shown for each of the five size classes according to equation (39). The base of reversal in settling direction, and the reversal layer thickness given by equations (41) and (42). (b) Mean concentration of particles ($\bar{N}$, particles/cm^3) as a function of depth. The curve shown is the sum total for the five size classes. Note very strong increase in $\bar{N}$ at the base of the reversal layers where each of the size classes falls out consecutively. Compare with equation (43). (c) Concentration of suspended matter ($\bar{S}$, µg/ ℓ). Note the difference between the concentrations expressed as $\bar{N}$ and $\bar{S}$.

addition of dissolved matter to ocean water in the process of particle dissolution (J, mass.vol.$^{-1}$time^{-1}) have been derived by *Lal and Lerman* [1973]. The main conclusions arising from the analysis of behavior of dissolving spherical particles are presented in Figure 2. Concentration of particles N increases very strongly in the

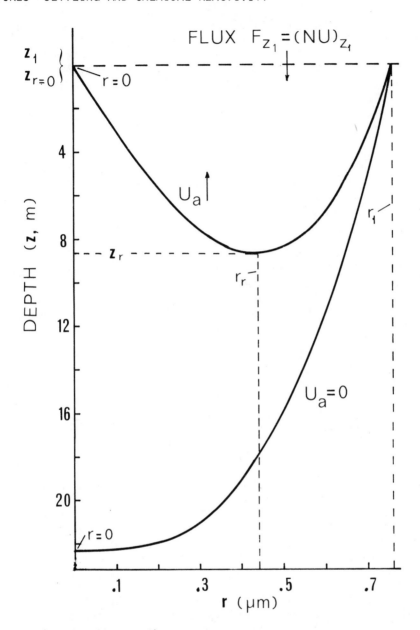

Fig. 3.   Radius of a dissolving spherical particle, as a
function of depth, in the presence U_a and in the absence ($U_a = 0$)
of upwelling velocity.   The settling flux F_{z_1} carries the particle
across depth z_1.   In the presence of upwelling velocity of water
U_a, the particles descend to depth z_r, where their settling direction is reversed, and they ascend until dissolution is complete at
$z_{r=0}$.   Equations (41) and (42).

reversal layers where particles of certain initial sizes ascend and dissolve to completion. Concentration of settling matter S, however, shows only very minor peaks (on the order of a few percent) at the depths of the reversal layers. This is owing to the fact that in $S = 4\pi\rho Nr^3/3$ the size of the particle radius has greater effect than the particle concentration N.

TRANSIENT PRODUCTION OF PARTICLES

Seasonal variations in the rates of nucleation and biological production of particles are important in the analysis of the particle behavior in shallow oceanic areas and in lakes.

Brunskill [1969, 1970; *Brunskill and Ludlam*, 1969] has conducted a series of two-year long observations on precipitation of calcite ($CaCO_3$) in a stratified lake, Green Lake, near Syracuse, New York. The essence of Brunskill's results pertinent to our treatment of suspended matter follows.   The upper water layer (approximately 20 m deep) of Green Lake is supersaturated with respect to calcite, and the ion activity product $IAP = a_{Ca^2+} \times a_{CO_3^{2-}}$ in it exceeds the saturation value for calcite (IAP_{eq}) by factors 2 to 8. The deeper part of the lake below the seasonal thermocline (depths between 20 m and 50 m; lake floor at 52 m) is also supersaturated with respect to calcite, but the degree of supersaturation is weaker: the values of the quotient IAP/IAP_{eq} are between 1.0 and 2.0. Calcite crystals form in the upper water layer during the months May through September. The crystals have rhombohedral and scalenohedral habit, and their mean volume in the deeper parts of the lake water column is equivalent to a sphere of radius $r \simeq 3$ μm. Figures 4 and 5 show the concentration/depth profiles of suspended calcite from May through September in 1966 and 1967.

A significant feature has been reported by the Green Lake investigators: despite the slight supersaturation (computed) in the deeper part of the water column, calcite crystals show signs of rounding and dissolution. Also, aggregates of crystals have been reported from the water column and lake floor sediments. The aggregates might have formed by adherence of crystals to mucous membranes of planktonic biota, and through ingestion by crustaceans [*Culver and Brunskill*, 1969].

While the variations in the near-surface concentration of calcite and the shape of the profiles may be related to changes in the rates of nucleation and physical characteristics of the shallow lake, the persistent occurrence of the concentration peak near 5 m depth for as long as 35-40 days in 1966 and 1967 suggests that nearly stationary conditions might have developed during those periods.

Origin of Concentration Peaks

While it is not the purpose of our paper to deal with the specific problem of calcite precipitation in Green Lake, we use *Brun-*

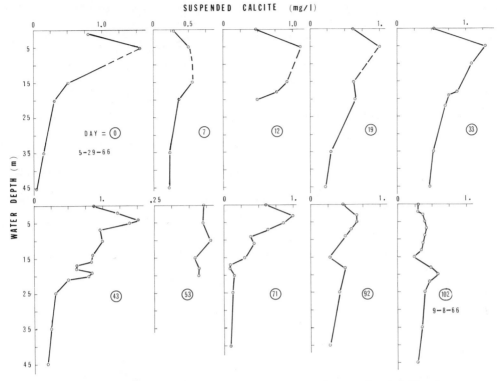

Fig. 4. Concentrations of suspended calcite (mg/ℓ) in Green Lake, New York, during the period May through September, 1966. From data in *Brunskill* [1970].

skill's [1969, 1970] data for demonstration of an analysis of the behavior of settling matter in situations similar to that of Green Lake.

The mechanisms of settling and dissolution of particles, responsible for the formation of concentration peaks in a water column (Figure 2), do not apply to a water column that is supersaturated with respect to the substance the particles are made of.

One possible explanation of the occurrence of transient peaks in concentration is variable production (nucleation) rates near water surface, as shown in Figure 6. Schematically, particles of three different sizes (r = 1, 1.5, and 2 μm) are being produced at different rates during the time interval T. Settling of these particles produces a concentration/depth profile at time T that is the sum of the concentrations of particles of different size. At time greater than T, the concentration peak between depths 2 and 3 in Figure 6 would spread out and split into three smaller peaks moving down and farther apart, owing to the differences between the settling velocities of the particles of different size. Persistence of the calcite concentration peaks near constant depth in Green Lake suggests that variable nucleation rates are not the main

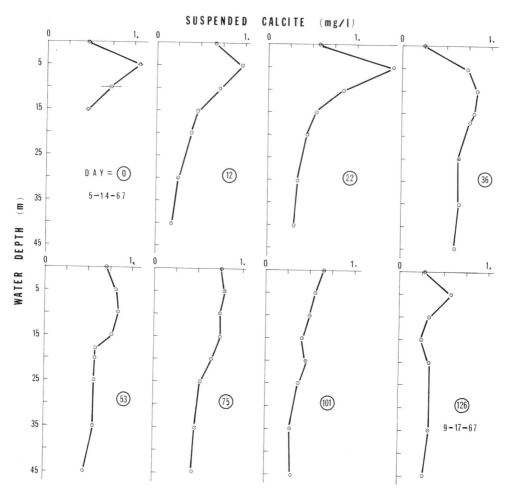

Fig. 5. Concentrations of suspended calcite (mg/ℓ) in Green Lake, New York, during the period May through September, 1967.

mechanism behind the peak formation.

The thermal structure of the lake water column, and supersaturation of near-surface water with respect to calcite suggest an alternative mechanism. The vertical temperature profile of the lake during its seasonal stratification (Figure 7) shows that water temperatures decrease from 21°- 25°C at the surface to 7°C at depth. The effect of this temperature change on water density (ρ, Figure 7) across the thermocline is negligible. A slight increase with depth in the concentration of dissolved matter also has a very small effect on lake water density [*Brunskill and Ludlam*, 1969]. The effect of temperature on viscosity of water, however, is significant: viscosity below the thermocline is 40-60% higher than in the surface water. Such an increase in viscosity produces an equivalent decrease in settling velocities, as shown in equation (2) and Fig-

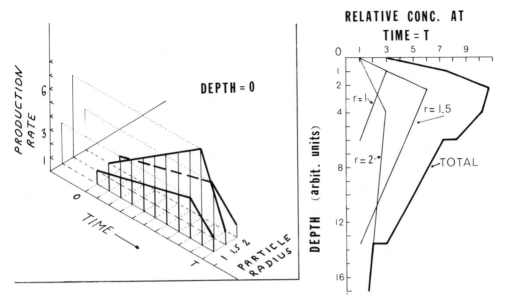

Fig. 6. Schematic representation of variable production rates of particles at the top of a water column (depth = 0).

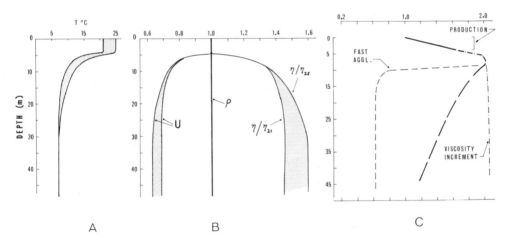

Fig. 7. (*a*) Range of vertical temperature variation in Green Lake during the period of summer stratification. From data in *Brunskill* [1970]. (*b*) Variations in density (ρ) and viscosity (η) of water, and the Stokes settling velocity of spherical particles *U* normalized to 1.0 at depth 0, owing to the temperature profile shown in (*a*). (*c*) Concentration of suspended particles, normalized to 1.0 at depth 0, obtainable under the following conditions: constant production rate in the upper 4 m of the water column, and settling into a higher viscosity. Agglomeration and/or dissolution of particles result in the mass concentrations decreasing with depth, as shown schematically by the other two curves.

ure 7. If the particles were produced at a steady rate per unit
volume in the upper 4 m of the water column, their settling through
the thermocline into the lower water layer of higher viscosity
would have produced a concentration/depth profile as shown in Figure 7, labelled "viscosity increment." (A more detailed account of
experimental work on settling velocities of spheres in a density
gradient has been given by *Kuo* [1972]).

If the increase in concentration of suspended calcite toward
the thermocline (Figures 4, 5) may be explained by a combination
of continuous production in the upper layer and increase in viscosity, an additional explanation should be given for the decrease in
concentration that is responsible for the occurrence of the peaks
on concentration/depth profiles.

One explanation is that if no particles nucleate below the
thermocline then the continuity of the particle flux requires that
the fluxes at any two depths must be equal,

$$(NU)_{z_1} = (NU)_{z_2} \quad , \tag{44}$$

where N is the particle concentration (particles.cm^{-3}), and U is
the Stokes settling velocity, and subscripts 1 and 2 refer to
depths z_1 and z_2. Concentration of suspended matter, made up of N
spherical particles, is $S = (4/3)\pi\rho_s r^3 N$ (g.cm^{-3}). Combining the
latter relationship with $U = Br^2$ in (44) gives

$$S_2/S_1 = r_2/r_1 \quad . \tag{45}$$

Thus, when the particle flux is conserved, a change in the concentration of suspended matter with depth is proportional to a change
in the particle radius. With reference to the data shown in Figures
4, 5, and 7, a decrease by a factor of two in S between two depths
(for example, 10 m and 45 m), should correspond to a decrease by a
factor of two in the particle radius, implying dissolution.

A mean rate of dissolution of a particle settling from z_1 to
z_2 can be written as

$$\bar{\varepsilon} = \frac{(r_1 - r_2)}{(z_2 - z_1)/\bar{U}} = \frac{(r_1-r_2)(r_1+r_2)^2 B}{4(z_2 - z_1)} \quad , \tag{46}$$

where $r = (r_1 + r_2)/2$ has been substituted in $\bar{U}$. For calcite spheres
in water at 7°C ($B = 2.65\times10^4$ cm^{-1}sec^{-1}, from $\rho_s - \rho = 1.7$ g.cm^{-3},
and $\eta = 0.014$ g.cm^{-1}sec^{-1}), mean rates of dissolution vary from
5×10^{-4}cm.yr^{-1} to 40×10^{-4}, for particles decreasing in radius from
2μm to 1μm, and from 4μm to 2μm, respectively, within the settling
distance of 35m. Such rates of dissolution are somewhat higher
than the experimental rates for sea water, discussed earlier in
this paper.

The second explanation of the concentration peaks is that, if
calcite particles neither dissolve nor grow but can form aggregates,

then the mass flux (as opposed to the particle flux) is conserved in the water column,

$$(SU)_1 = (SU)_2 \quad . \tag{47}$$

Using the notation of the preceding paragraphs, (47) reduces to

$$S_2/S_1 = (r_1/r_2)^2 \quad , \tag{48}$$

which indicates that if the concentration of suspended matter (mass/volume) decreases with depth, the particle size should increase. With reference to the Green Lake data, a decrease in S by a factor of two between depths 10 m and 45 m corresponds to an increase in particle size by a factor $r_2/r_1 = 2^{\frac{1}{2}} = 1.41$. Such an increase in the Stokes settling radius can be achieved through formation of three-particle aggregates (from simple geometric considerations, the radius of a sphere equal in volume to three smaller spheres, each of radius r, is $3^{1/3}r = 1.44r$).

The preceding discussion of the behavior of reacting particles in a physically inhomogeneous water column indicates the kind of data that are needed for a theoretical treatment of problems, as well as the general difficulties involved in the use of deterministic models. An alternative stochastic model, applicable to a broader class of problems, is presented in the subsequent section.

STOCHASTIC MODEL OF PARTICLE TRANSPORT

The model described here is a modification of a random walk model. Details of mathematical derivation will be published elsewhere [Dacey, in preparation, 1973], while here we present the underlying principles and operational techniques aimed at the use of the model.

A column of water shown in Figure 8 may be thought of as subdivided into a series of cells or stages, each of length L. Particles enter the first cell at a specified rate, move through the water column, and exit from the last cell without returning. A particle that is in the jth cell at time t can occur at time $t + \Delta t$ in any one of the several other cells of the water column, and the probability of its transition from the jth cell to any other cell may vary with the other cell's location. In the model shown in Figure 8, a particle in the upper part of the water column always moves from cell j to cell $j + 1$ within some specified time interval Δt. This corresponds to the transition probability of $P(j, j+1) = 1$. Further down the water column, the settling velocities of particles decrease. In this region of the water column, in a time interval Δt, a particle may either remain within the jth cell, move downward to the cell $j + 1$ or move upward to cell $j - 1$. For this region, the transition probabilities are $P(j, j-1)$, $P(j, j)$ and $P(j, j+1)$ and are subject to the condition

$$P(j, j-1) + P(j, j) + P(j, j+1) = 1. \tag{49}$$

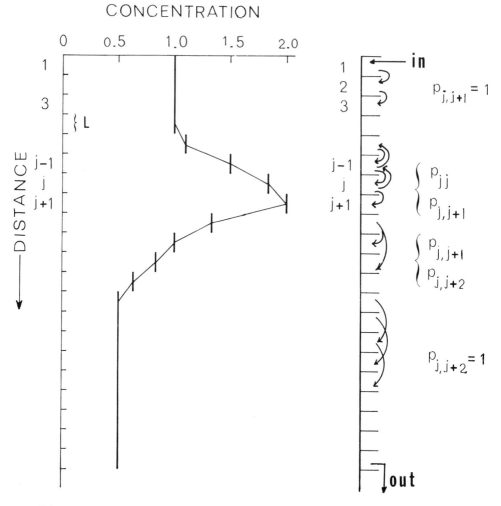

Fig. 8. Stochastic model of particle settling. On the right, water column subdivided into cells, each of length L. Particles enter at the top and exit at the bottom. Their transport through the column is controlled by transition probabilities P_{ij}. On the left, concentration profile computed using equation (49) and the rate of particle input to cell 1 taken as $G_t = 1$ particle per unit time step. Note the relationships between the relative concentration values shown and the values of transition probabilities P_{ij} in matrix P.

Further down the water column, the settling velocities of the particles increase such that, for example, a particle moves the distance $2L$ during a time interval Δt. In this region, the transition probabilities are $P(j, j+2) = 1$.

The matrix P, shown below, contains the transition probabili-

MATRIX P FOR FIGURE 8

	1	2	3	4	5	6	7	8	9	10	11	12	13	14	15	16	17	18	19	20	21	22
1	1																					
2		1																				
3			1																			
4				1																		
5				.1	.9																	
6					.25	.75																
7						.45	.55															
8							.5	.5														
9								.2	.8													
10										1												
11										.8	.2											
12											.4	.6										
13												.1	.9									
14														1								
15															1							
16																1						
17																	1					
18																		1				
19																			1			
20																				1		
21																					1	
22																						1

ties $P(i,j)$, where i, $j = 1, \ldots, J, J+1$, and J is the number of cells in the water column. All the transition probabilities $P(i, j)$ other than those shown are zeros. The entry $P(J+1, J+1) = 1$ is included in the matrix as a condition that assures no return of particles to the water column. The water column in Figure 8 is subdivided into $J = 21$ cells; matrix P is accordingly of size 22 x 22.

By varying the structure of the transition probability matrix P, concentration maxima and minima of different magnitudes can be obtained. To construct a transition probability matrix P for a particular water column, one must have a certain amount of information on those characteristics of the system that affect the settling behavior of particles. For example, to obtain a concentration/depth profile of zero gradient, as in the upper part of the water column in Figure 8, the unit cell length L and unit time step Δt may be chosen such that the settling velocity is $U = L/\Delta t$. Then the transition probabilities can be defined as $P(j, j+1) = U\Delta t/L = 1$. If the settling velocity becomes smaller farther down in the water column, then $P(j, j+1) < 1$ and $P(j, j) = 1 - P(j, j+1)$. Similarly, for greater values of the settling velocity transition probabilities of the type $P(j, j+2)$ are greater than 0.

The physical meaning of transition probabilities of the type $P(j, j-1)$ is that particles move upward. One possible mechanism of upward movement is entrapment of particles in vertical convective

cells [*Stommel*, 1949] of dimensions greater than L (Figure 8).
Another possible mechanism is dissolution of particles in an upwelling regional mass of water, treated earlier in this paper.

If the rate of input to cell 1 is G_t particles per time interval Δt, then the expected number of particles at time t in cell
$j(t \geq j)$, denoted by $\underline{EN}_t(j)$, is [*Dacey*, 1973]

$$\underline{EN}_t(j) = \sum_{k=j-1}^{t-1} G_{t-k} P^k(1, j) , \tag{50}$$

where time t is an integer and multiple of the unit time step Δt,
P^k is the kth power of the transition probability matrix P, and
$P^k(i, j)$ is the (i, j) entry in P^k. For example, the expected
number of particles in cell $j = 3$ at time $t = 6$ is, from (50), a sum
of four terms

$$\underline{EN}_6(3) = G_4 P^2(1, 3) + G_3 P^3(1, 3) + G_2 P^4(1, 3) + G_1 P^5(1, 3),$$

where the rates of input to cell 1, G_t, at consecutive times may
differ from one time to another.

Raising the transition probability matrix P to consecutive
powers, and summation of the terms in (50) is easily and inexpensively
accomplished on a digital computer.

In the case of a steady input rate to cell 1 of one particle
per unit time step so that each $G_t = 1$, a steady-state concentration
of particles in the water column can be obtained from the following
procedure which is simpler than the summation of terms in (50) up
to large values of time t [*Dacey*, 1973]. Omit the last row and
last column in matrix P, to obtain a $J \times J$ matrix P'. Next, define
matrix Q which is the inverse of the difference between unit matrix I
($J \times J$, $I(j, j) = 1$ and $I(i, j) = 0$ for all $i \neq j$) and matrix P':

$$Q = (I - P')^{-1}. \tag{51}$$

The elements of the first row of matrix Q, $Q(1, j)$ for $j = 1, \ldots,$
J, are the expected values of the number $N(j)$ of particles in cell
j when $t \to \infty$:

$$\underline{EN}(j) = Q(1, j). \tag{52}$$

The above procedure for obtaining the values of $\underline{EN}(j)$ involves only
one matrix inversion. Figure 8 shows the values $\overline{EN}(j)$ when the
transition matrix P is of the above form.

A case of practical importance is input of particles to a
water column at more than one depth. Such a situation may arise
when particles are produced throughout a long section of a water
column or when horizontal advective input is significant. A

computational expression for handling this type of situation is illustrated in order to demonstrate the inherent versatility of the basic model.

Suppose that at each unit time one particle is input to the water column, and the probability that it enters at cell i is $\pi(i)$, where $\pi(1) + \pi(2) + \ldots + \pi(J) = 1$. Then it may be shown that

$$\underline{E}\, N_t(j) = \sum_{k=0}^{t} \sum_{i=1}^{J} P^k(i,j)\pi(i), \qquad (53)$$

which is a convenient form for computer analysis. Further, the procedure leading to (52) may be modified to obtain

$$\underline{E}\, N(j) = \sum_{i=1}^{J} Q(i,j)\pi(i). \qquad (54)$$

SUMMARY AND CONCLUSIONS

The Stokes settling velocities of disc- and cap-shaped particles are about 2 times lower than the settling velocity of a sphere of the same volume and mass. The computed settling velocities for cylindrical particles, however, are between 5 and 25 times higher than for a sphere of the same volume and mass, depending on the sphere size (Figure 1).

If the particles of different shapes (Figure 1) dissolve in the course of settling according to the constant-rate mechanism (i.e., the rate of mass loss ε', in units of $g.cm^{-2}yr^{-1}$, is constant), then their longevities in the water column, as well as their contributions to dissolved matter in ocean water, are functions of their surface areas. Comparing the different shapes with the sphere, the surface areas can be written in terms of the sphere radius r and other parameters, identified in Table 1 and Figure 1, in a manner similar to the particle volumes. When this is done for particles of the same volume ($4\pi r^3/3$) and dimensions given in Figure 1, one finds that the surface areas of the two spheroids of revolution exceed the surface area of the sphere by less than 10%. Since the settling velocities of the spheroids and the sphere differ by about 10-20% (Figure 1), there is no pronounced effect of shape on the rate of dissolution and settling.

The two cylinders of dimensions as shown in Figure 1 have surface areas 20% (circular) and 40% (elliptic) greater than the sphere of the same volume. Thus, dissolution would not appreciably affect their settling velocity by comparison with the dissolution effect on settling of the sphere.

The surface areas of the disc and hemispherical cap, of dimensions given in Figure 1, are 3.1 and 3.6 times greater than the sphere surface area. The relatively large surface areas, coupled with the slow settling rates, may be responsible for shorter settling distances, all other factors being equal.

Theoretical treatment of the Stokes settling of dissolving particles indicates a number of possible shapes of concentration/depth profiles, as shown in Figures 2 and 3. To evaluate the *in situ* rates of particle dissolution or growth, one must know (*i*) concentrations of particles, (*ii*) their mineralogical nature, and (*iii*) size distributions at different depths. Such information would be sufficient for the first test and further development of the model of the behavior of non-living suspended matter, as presented in the section containing equations (38)-(43).

Transient phenomena in the production and settling of particles in a water column are amenable to treatment according to the stochastic model given by equations (49) - (54). The value of this model is its ability to predict and reproduce concentrations of particles in a physically heterogeneous water column and under variable conditions of particle input. Use of the model requires knowledge of the physical and chemical characteristics of the system, such as the thermal structure of the water column, degree of turbulence, modes of change in particle size, and the structure of input, all of which may vary with the depth thereby affecting the values of transition probabilities that describe transport of a particle through the water column.

ACKNOWLEDGMENTS

Dr. G. J. Brunskill, of the Fisheries Research Board of Canada, Winnipeg, Manitoba, made available to us his unpublished data on suspended calcite in lake water. The research was supported by the Oceanography Section, National Science Foundation, NSF Grant GA-30769.

REFERENCES

Allen, T., *Particle Size Measurement*, Chapman and Hall, London, 1968.

Berger, W. H., Foraminiferal ooze: solution at depths, *Science*, *156*, 383-385, 1967.

Berger, W. H., Radiolarian skeletons: solution at depths, *Science*, *159*, 1237-1239, 1968.

Berner, R. A., *Principles of Chemical Sedimentology*, McGraw-Hill, New York, 1971.

Biscaye, P. E., Mineralogy and sedimentation of Recent deep-sea clay in the Atlantic Ocean and adjacent seas and oceans, *Bull. Geol. Soc. America*, *76*, 803-832, 1965.

Breach, D. R., Slow flow past ellipsoids of revolution, *J. Fluid Mech.*, *10*, 306-314, 1961.

Brenner, H., The Stokes resistance of a slightly deformed sphere, *Chem. Eng. Sci.*, *19*, 519-539, 1964.

Brunskill, G. J., Fayetteville Green Lake, New York. II. Precipitation and sedimentation of calcite in a meromictic lake with laminated sediments, *Limol. Oceanogr.*, *14*, 830-847, 1969.

Brunskill, G. J., *Supplementary Physical and Chemical Data for Fayetteville Green Lake, N. Y.*, Freshwater Institute, Fisheries Research Board of Canada, Winnipeg, Man., 55 p., mimeogr., 1970.

Brunskill, G. J., and S. D. Ludlam, Fayetteville Green Lake, New York. I. Physical and chemical limnology, *Limnol. Oceanogr.*, *14*, 817-829, 1969.

Chester, R., Geological, geochemical and environmental implications of the marine dust veil, in *The Changing Chemistry of the Oceans, Proc. 20th Nobel Symposium*, edited by D. Dyrssen and D. Jagner, Wiley Interscience, New York, 291-305, 1972.

Craig, H., Abyssal carbon and radiocarbon in the Pacific, *J. Geophys. Res.*, *74*, 5491-5506, 1969.

Culver, D. A., and G. J. Brunskill, Fayetteville Green Lake, New York. V. Studies of primary production and zooplankton in a meromictic marl lake, *Limnol. Oceanogr.*, *14*, 862-873, 1969.

Dacey, M. F., Recurring random walk model for sediment transport, in preparation, 1973.

Garrels, R. M., and R. M. Dryer, Mechanism of limestone replacement at low temperatures and pressures, *Bull. Geol. Soc. America*, *63*, 325-379, 1952.

Gibbs, R. J., The geochemistry of the Amazon River system, *Bull. Geol. Soc. America*, *78*, 1203-1232, 1967.

Gibbs, R. J., M. D. Matthews, and D. A. Link, The relationship between sphere size and settling velocity, *J. Sed. Petrol.*, *41*, 7-18, 1971.

Goldberg, E. D., Minor elements in sea water, in *Chemical Oceanography*, vol. 1, edited by J. P. Riley and G. Skirrow, 163-196, Academic Press, New York, 1965.

Hurd, D. C., Factors affecting solution rate of biogenic opal in sea water, *Earth Planet. Sci. Lett.*, *15*, 411-417, 1972.

Hutchinson, G. E., *A Treatise on Limnology*, vol. 2, Wiley, New York, 1967.

Jacobs, M. B., and M. Ewing, Mineralogy of particulate matter suspended in sea water, *Science, 149*, 179-180, 1965.

Krey, J., Detritus in the ocean and adjacent sea, in *Estuaries, Amer. Assoc. Adv. Sci. Pub. 83*, edited by G. H. Lauff, pp. 389-394, Amer. Assoc. Adv. Sci., Washington, D. C., 1967.

Kuo, C. Y., *Free Falling Particle in Density Stratified Fluid, Completion Report A-032-PR*, Water Resources Research Institute, University of Puerto Rico, Mayagüez, P. R., 48 pp., mimeogr., 1972.

Lal, D., and A. Lerman, Dissolution and behavior of particulate biogenic matter in the ocean: some theoretical considerations, *J. Geophys. Res., 78*, 7100-7111, 1973.

Lamb, H., *Hydrodynamics*, 6th edition, Dover, New York, 1932 (1945).

Lerman, A., Time to chemical steady states in lakes and ocean, *Adv. Chem. Ser., 106*, 30-76, 1971.

Lerman, A., F. T. Mackenzie, and L. B. Plummer, Mineral dissolution and precipitation: S-shaped kinetics, *Eos Trans. AGU, 54*(4), 341, 1973.

Levich, V. G., *Physicochemical Hydrodynamics*, Prentice-Hall, Englewood Cliffs, N. J., 1962.

Manheim, F. T., R. H. Meade, and G. C. Bond, Suspended matter in surface waters of the Atlantic continental margin from Cape Cod to the Florida Keys, *Science, 167*, 371-376, 1970.

Morse, J. W., and R. A. Berner, Dissolution kinetics of calcium carbonate in sea water: II. A kinetic origin for the lysocline, *Amer. J. Sci., 272*, 840-851, 1972.

Munk, W. H., Abyssal recipes, *Deep Sea Res., 13*, 707-730, 1966.

Munk, W. H., and G. A. Riley, Absorption of nutrients by aquatic plants, *J. Mar. Res., 11*, 215-240, 1952.

Nielsen, A. E., *Kinetics of Precipitation*, Macmillan, New York, 1964.

Pasquill, F., *Atmospheric Diffusion*, Van Nostrand, New York, 1962.

Payne, L. E., and W. H. Pell, The Stokes flow problem for a class of axially symmetric bodies, *J. Fluid Mech., 7*, 529-549, 1960.

Peterson, M. N. A., Calcite: rates of dissolution in a vertical
    profile in the Central Pacific, *Science*, *154*, 1542-1544, 1966.

Plummer, L. N., *Rates of Mineral-Aqueous Solution Reactions*, Ph. D.
    thesis, Dept. Geological Sciences, Northwestern Univ., Evanston,
    Ill., 1972.

Redfield, A. C., B. H. Ketchum, and F. A. Richards, The influence
    of organisms on the composition of sea-water, in *The Sea*,
    vol. 2, edited by M. N. Hill, pp. 26-77, Wiley Interscience,
    New York, 1963.

Riley, G. A., Particulate and organic matter in sea water, *Adv.
    Mar. Biol.*, *8*, 1-118, 1970.

Shutz, D. F., and K. K. Turekian, The investigation of geographical
    and vertical distribution of several trace elements in sea
    water using neutron activation analysis, *Geochim. Cosmochim.
    Acta*, *29*, 259-313, 1965.

Smith, F. B., The turbulent spread of a falling cluster, *Adv.
    Geophys.*, *6*, 193-210, 1959.

Somayajulu, B. L. K., D. Lal, and S. Kusumgar, Man-made carbon-14
    in deep Pacific waters: transport by biological skeletal
    material, *Science*, *166*, 1397-1399, 1969.

Sorokin, Y. I., Microbial activity as a biogeochemical factor in
    the ocean, in *The Changing Chemistry of the Oceans, Proc.
    20th Nobel Symposium*, edited by D. Dyrssen and D. Jagner,
    189-204, Wiley Interscience, New York, 1972.

Stommel, H., Trajectories of small bodies sinking slowly through
    convection cells, *J. Mar. Res.*, *8*, 24-29, 1949.

Taylor, T.D., Low Reynolds number flows, in *Basic Developments in
    Fluid Dynamics*, vol. 2, edited by M. Holt, pp. 183-215, 1968.

Turekian, K.K., A. Katz, and L. Chan, Trace element trapping in
    pteropod tests, *Limnol. Oceanogr.*, 18, 240-249, 1973.

Weber, W.J., *Physicochemical Processes for Water Quality Control*,
    Wiley Interscience, New York, 1972.

Williams, P. M., J. A. McGowan, and M. Stuiver, Bomb ^{14}C in deep
    sea organisms, *Nature*, *227*, 375-376, 1970.

Wollast, R., Kinetics of the alteration of K-feldspar in buffered
    solutions at low temperature, *Geochim. Cosmochim. Acta*, *31*,
    635-648, 1967.

II

PRINCIPLES OF OPTICAL TECHNIQUES

Beam Transmissometers for Oceanographic Measurements

JOHN E. TYLER, ROSWELL W. AUSTIN, AND
THEODORE J. PETZOLD

Scripps Institution of Oceanography

ABSTRACT

*The theoretical basis for the measurement of the beam trans-
mittance of water is reviewed. Existing data for the beam transmittance of clean and distilled water are compared to demonstrate
the current state of knowledge regarding the beam transmittance of
clean water. Several recent instruments of sophisticated design are
briefly described and data are given for surface particle concentration.*

INTRODUCTION

The importance of optical measurements for studying the properties of particulate matter in the ocean has been demonstrated at
both the theoretical and experimental levels. At the present time
the theoreticians seem to have progressed further than have the
experimentalists. This is due, perhaps, to the fact that the former
enjoy certain simplifications not found in the real world; for example, they are able to work with spherical particles which, con-
veniently, refrain from multiple scattering. Further, the theoreticians never have the problem of an adequate and stable supply of
radiant flux; there is never any time variability in the composition
of their samples; and the measurement concepts they employ do not
engage in mutual interference. The experimentalist, on the other
hand, must deal with the real ocean and with the complex optical
problems associated with its real burden of particulate matter. His
job is to devise the instruments and measure the optical quantities
defined by theory.

Each theoretically defined optical quantity confronts the experimentalist with unique requirements for the control of the radiant flux involved. There is almost always a gross mismatch between the spectral and geometrical emission properties of the available source of radiant flux and the properties required, by definition, for the measurement. Radiant flux that does not conform to a definition from theory becomes spurious and must be eliminated. Unfortunately, the techniques for eliminating spurious radiant flux cannot be systematically organized, the procedures being more a matter of recognizing the problem, locating the source of error, and providing a solution. Failure to eliminate spurious radiant flux easily leads to gross errors.

BEAM TRANSMISSOMETERS

The idea of measuring the transmittance of water by means of a beam of light has been used (and abused) for many years. The beam transmittance for exceptionally clean water is of special interest since it represents a limiting case. In Figure 1, four of the most widely referenced determinations of the total attenuation coefficient for exceptionally clean water are compared. These data have been computed directly from measurements of the beam transmission over long paths.

It is interesting that *Clarke and James* [1939] and *James and Berge* [1938] used identical equipment in both studies with only one modification: Clarke and James lined the sample tube with ceresin wax, whereas James and Berge used a bright silver lining in the tube. The implication of the data from the two studies is consequently obvious. The silver lining acted to reflect much more of the forward scattered light toward the photodetector resulting in apparent high values of transmittance which, in turn, yielded low and incorrect values for the attenuation coefficient in the blue region of the spectrum where absorption is low. In the red region of the spectrum the high absorption of water significantly reduced this effect.

Sullivan [1963] and *Hulburt* [1945] both used glass tubes with no internal baffles. In both of these cases there exists the very real possibility that forward scattered light could have been reflected toward the photodetector by total internal reflection from the walls of the glass tubing and, further, that the reported attenuation coefficients in the blue region of the spectrum are still too small for clean water.

In the phenomological theory of radiative transfer developed by *Preisendorfer* [1960], the equation of transfer is given in the form,

$$\frac{1}{V} \frac{d[N/n^2]}{dr} = -(a + s) \frac{N}{n^2} . \tag{1}$$

where

$$a \frac{N}{n^2} \quad \text{is the absorption-loss term}$$

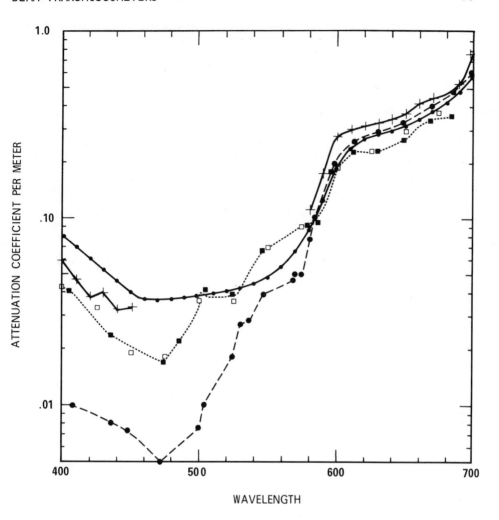

Fig. 1. Spectral values of the total attenuation coefficient for exceptionally clean water, computed from measurements of beam transmittance. (· — ·) = data of *Hulburt* [1945]; (+ — +) = data of *Sullivan* [1963]; (□ ···· □) = data of *Clarke and James* [1939]; (□) = attributed to Clarke and James by *Jerlov* [1968]; (· -- ·) = data of *James and Berge* [1938].

and

$$s \frac{N}{n^2}$$ is the scattering-loss term.

Preisendorfer points out that the scattering mechanism which causes the scattering loss simultaneously increases the population of the packet of photons by causing photons to be scattered into the direction of travel of the radiance N. To account for this, he alters (1) to give,

$$\frac{1}{V} \frac{d[N/n^2]}{dr} = -(a + s) \frac{N}{n^2} + \frac{N^\star}{n^2} , \tag{2}$$

where $N^\star$ represents the radiance per unit path length due to photons which, after being scattered, travel in the direction of the radiance N.

The sum $a + s$ is set to equal a and represents losses from the direction of travel, due to absorption a and scattering s.

It is next assumed that the light field is in a steady state, that the ocean is emission-free and isotropic, and that the index of refraction underwater remains constant. On this basis, Preisendorfer rewrites the equation of transfer in the familiar form,

$$\frac{dN}{dr} = -aN + N^\star \tag{3}$$

which, when solved for a, yields,

$$a = \frac{N^\star}{N} - \frac{1}{N} \cdot \frac{dN}{dr} . \tag{4}$$

Experimental procedures for determining the value of a are dependent on optical techniques which attempt to make one or the other of the two terms on the right-hand side of (4) negligible or zero.

If $\frac{N^\star}{N}$ becomes 0, then

$$a = -\frac{1}{N} \frac{dN}{dr} , \tag{5}$$

which is the basic equation for the beam-transmissometer method for determining a.

If $\frac{1}{N} \frac{dN}{dr}$ becomes 0, then

$$a = \frac{N^\star}{N} , \tag{6}$$

which is the basic equation for the black-target technique for determining a.

For *in situ* determinations of a, measurements with a beam transmissometer are, by far, the most often used. In a practical beam transmissometer, there will be spurious flux which comes from small-angle forward scattering that is within the collection capability of the detector. This source of spurious radiant flux was recognized by *Wills and Jones* [1953] at the Admiralty Research Laboratory, England. They estimated that, for their instrument, the average error under adverse conditions of measurement could be as high as 34% in the value of the attenuation coefficient. More re-

cently, *Latimer* [1972] used Mie theory to compute the extinction
efficiency as a function of *Van de Hulst's* [1957] particle-size
parameter for monodispersed spherical particles. His computations
show that this particular source of spurious radiant flux will
decrease significantly as the ratio of beam diameter to beam length
becomes smaller.

   Priesendorfer [1958, 1960] investigated the relationship between
the error introduced by the dimensions of the light beam used in the
transmissometer. This relationship, shown in Figure 2, indicates
the advantage gained by using a non-diverging beam of light having
a small ratio of beam diameter to beam length.

   There are, of course, other potential sources of spurious flux
in beam-transmissometer measurements. One such source is the ambient
natural daylight which may interfere either during measurement in
air or during measurement in water.

   Since the determination of α requires a measurement in air as
well as a measurement in water, it is necessary to provide an optical system that will insure that all of the projected flux will
arrive at the detector under both conditions. In effect, this
means that all rays within the beam which are refracted as a result
of immersion in water must be controlled so that they impinge on the
photodetector-sensitive surface both before and after refraction.
This can be accomplished by making the image of the field stop of the
projector at the photodetector the same size as the aperture stop
of the projector. All rays will then be confined in the cylinder
defined by the projector aperture at the projector end and the image
of the field stop at the receiver end, and all rays affected by
refraction in water will remain on the detector-sensitive surface.

   The windows for the watertight cases of a beam transmissometer

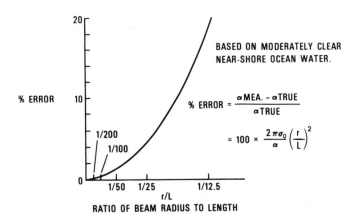

Fig. 2. Theoretical relationship between the error due to
collection of spurious forward-scattered light and the dimensions
of the transmissometer beam.

must be carefully designed to prevent their becoming a detrimental
factor. They must not be a source of scattered light and they
must not distort the structure of the beam of light when under
pressure due to submergence. Since the interface reflectance of
the exposed window surfaces will be different in water than in air,
it is necessary to know the index of refraction of the glass in
order to make the necessary correction. Failure to make this
correction will result in transmittance values greater than 100%
per meter for some clear waters.

Spectral values of the total attenuation coefficient are of
great interest but impose further measurement problems since
any radiant flux outside a desired bandwidth of wavelengths
becomes, by definition, spurious flux and must be eliminated.
The magnitude of these several sources of spurious flux can be
influenced by the length of the water path being measured and by
the magnitude of the optical properties themselves. Hence it is
important to have supporting measurements of as many other optical
properties as possible. *Petzold and Austin* [1968] have incorporated in
beam transmissometers design features which avoid, to a considerable
extent, the spurious flux problems outlined above. An optical
schematic of their 1-m folded-path transmissometer is shown in
Figure 3. The light beam for this instrument has a 20-mm diameter
and a length of 1 m, yielding a ratio of r/L of 1/100 with an
estimated error from this source of less than 1%. The projector
and photodetector are housed in a single watertight case and the
projected beam of light is folded back by means of a poroprism
as shown in Figure 3. The shutter and light pipe which will,
on command, direct a fraction of the lamp output to the photodetector for the purpose of monitoring the stability of the source
are also shown in Figure 3.

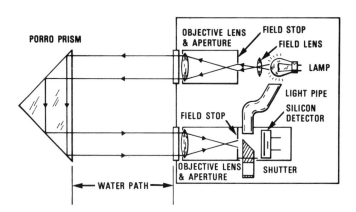

Fig. 3. Schematic diagram of the folded-path transmissometer
shown in Figure 4.

The projection system of the transmissometer produces a
cylindrically restricted beam of light which confines all projected
rays within a cylinder defined by the aperture of the projector
lens and the image of the projector field stop which is focused
on the receiver system in such a way that all these rays fall
on the sensitive surface of the photodetector. These rays will
remain on the photodetector when the index of refraction along
the path changes from that of air to that of water.

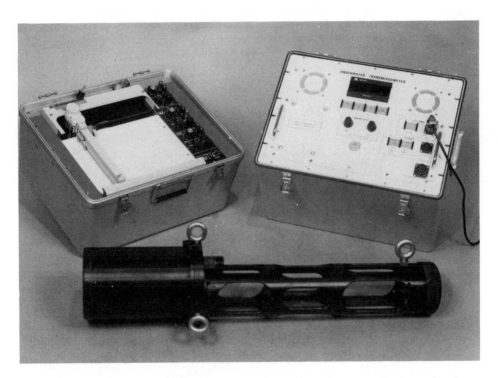

Fig. 4.   Beam transmissometer with 1-m folded path for single-
filter operation. The x-y recorder plots transmittance and
temperature as a function of depth.

Figure 4 shows a recent model of the beam transmissometer
having a 1-m path. This instrument uses a fixed spectral
bandwidth with automatic recording of both beam transmittance and
temperature as a function of depth. An option to this instrument
is available with five spectral bandwidths which can be selected
by remote control. Both instruments were designed for optical
studies in the euphotic zone and have a depth capability of 300 m.
      A beam transmissometer [*Hughes and Austin,* 1965] having an
adjustable water path of from 1/2 to 2 m and a depth capability

of 2000 m which makes use of a rotating sector disc and neutral
density wedge for null-balance operation has been used. A
schematic of the optical system is shown in Figure 5.

For continuous measurements of beam transmittance to 5000 m,
it would be a simple matter to incorporate the favorable design
features of these existing transmissometers in a self-contained
free-falling instrument that would require no cable. Continuous
monitoring or automatic control of the lamp output could be employed to maintain accuracy in data reduction. The instrument could
also be provided with a separate detector arranged to sense the
particle scattering at a single angle in the forward direction.

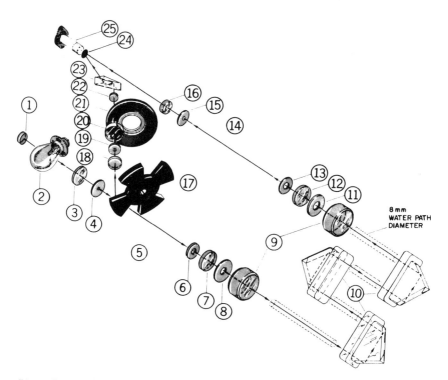

Fig. 5. Schematic optical system of the null-balance beam
transmissometer of *Hughes and Dustin* [1965]. (1)-(8) comprise
its projection system; (9)= housing windows; (10)= the prism system for folding the path length; (11)-(16)= the detector optical
system; (17)= a rotating sector disc which, by means of (3), (18),
(19), (20), (22), and (23) intermittently images the field
stop (4) on the photodetector (25); (24)= a spectral filter;
(21)= a neutral density wedge which, by means of a servo mechanism,
equates the two signals, one via the water path and the other
via the air part within the instrument housing.

REFERENCES

Clarke, G. L. and H. R. James, Laboratory analysis of the selective absorption of light by sea water, *J. Opt. Soc. Am.*, *29*, 43-55, 1939.

Hughes, R. S. and R. W. Austin, Deep-Sea Light Attenuation Measurements With a Null-Balance Transmissometer. *Tech. Pub. 3748*, U. S. Naval Ordnance Test Station, China Lake, 1965.

Hulburt, E. O., Optics of distilled and natural water, *J. Opt. Soc. Am.*, *35*, 698-705, 1945.

James, H. R. and E. A. Birge, A Laboratory Study of the Absorption of Light by Lake Waters, *Trans. Wisc. Acad. Sci.*, *Arts, and Letters*, *31*, 154 pp., 1938.

Jerlov, N. G., *Optical Oceanography*. Elsevier, London, 194 pp., 1968.

Latimer, I., Dependence of Extinction Efficiency of Spherical Scatterers of Photometer Geometry, *J. Opt. Soc. Am.*, *62*, 208-211, 1972.

Petzold, T. R. and R. W. Austin, An Underwater Transmissometer For Ocean Survey Work, *Scripps Institution of Oceanography*, *Ref. 68-9*, 5 pp., 1968.

Preisendorfer, R. W., A General Theory of Perturbed Light Fields, With Applications to Forward Scattering Effects in Beam Transmittance Measurements, *Scripps Institution of Oceanography*, *Ref. 58-37*, 19 pp., 1958.

Preisendorfer, R. W., Application of radiative transfer theory to light measurements in the sea, *Symposium on Radiant-Energy in the Sea*, pp. 11-30, International Union of Geodesy and Geophysics, Paris, 1960.

Sullivan, S. A., Experimental study of the absorption in distilled water, artificial sea water, and heavy water, in the visible region of the spectrum, *J. Opt. Soc. Am.*, *53*, 962-968, 1963.

van de Hulst, H. C., *Light Scattering By Small Particles*, New York, John Wiley & Sons, New York, 1957.

Wills, M. S. and D. Jones, The Effect of Forward Scattered Light on the Readings of a Light Transmission Meter, *Report A. RL/RI/K301*, Admiralty Research Laboratory, Teddington, Middlesex, England, 12 pp., 1953.

Volume-Scattering Functions in Ocean Waters

RAYMOND C. SMITH, ROSWELL W. AUSTIN, AND
THEODORE J. PETZOLD

Scripps Institution of Oceanography

ABSTRACT

Absolute values of the volume-scattering function, obtained in situ for a variety of natural ocean waters, are presented. The use of optical techniques to infer the concentration, optical characteristics, and size distribution of suspended particulate material in the ocean is outlined. The technique of using single-angle scatter meters to estimate the total volume-scattering function (and hence the suspended particle concentration) is discussed and an optimization of the technique, based upon the data presented, is suggested.

INTRODUCTION

The characteristics of suspended particulate material are related to basic biological, chemical and physical processes in the oceans. As a consequence, a careful study of the properties and distribution of suspended particles can provide valuable information about the oceanographic processes affecting these particles. A challenge to the use of optical techniques as an oceanographic tool is the extraction of characteristic information about the scattering particles (e.g., concentration, size distribution, index of refraction, refractive index distribution, shape parameter) from appropriate optical measurements. *In situ* optical techniques for investigating suspended particles appear to be particularly advantageous when the particles are not readily accessible, when particle sampling techniques alter the *in situ* characteristics, and when numerous measurements are necessary for wide spatial and temporal coverage.

In spite of these apparent advantages the full potential of utilizing optical techniques as a tool for *in situ* particle analysis for the study of basic oceanographic processes has not yet been utilized. This is due both to the theoretical difficulty of characterizing a polydispersed distribution of particles from measured optical properties and to the experimental difficulties of completely and accurately measuring the *in situ* optical properties.

The theoretical complexity is made tractable by an apparent quasi-stationary distribution of particle sizes in the upper layers of the ocean [*Bader*, 1970; *Carder et al.*, 1971; *Owen*, 1972] and by appropriate use of simplifying assumptions and large computers [*Kullenberg*, 1970; *Kullenberg and Olsen*, 1972; *Gordon and Brown*, 1972]. The full experimental difficulties have frequently been avoided by making use of the observation [*Jerlov*, 1953; *Tyler*, 1961; *Kullenberg*, 1968; *Beardsley et al.*, 1970] that a linear correlation exists between the volume-scattering function measured at $\beta(45°)$ and the total scattering coefficient b. Thus, an estimate of the total scattering coefficient can be obtained from a measurement of $\beta(45°)$. However, these same observations show that the ratio of $\beta(45°)$ to b varies by a factor of two or three for different water masses.

Accurate and complete values of the volume-scattering function [*Petzold*, 1972], obtained *in situ* for a variety of natural waters, allow a reconsideration of past strategies for inferring the characteristics of suspended particles from optical measurements. In particular, utilizing these recent results, the method by which the accuracy of estimating the total volume-scattering function from a single-angle scatter meter can be optimized will be shown below.

THE VOLUME-SCATTERING AND PHASE FUNCTIONS

The volume-scattering function [*Tyler and Priesendorfer*, 1962] is defined as

$$\beta(\Theta) = \frac{dI(\Theta)}{E \cdot dV} \tag{1}$$

Here $dI(\Theta)$ is the radiant intensity scattered from a small volume dV in a direction Θ from the initial direction for an irradiance E incident on the volume. The total scattering coefficient b is found by integration over all solid angles $d\Omega = 2\pi \sin d\Theta$,

$$b = \int_{4\pi} \beta(\Theta)d\Omega = 2\pi \int_0^\pi \beta(\Theta) \sin \Theta \, d\Theta . \tag{2}$$

Using Mie's solution [*Van de Hulst*, 1957] for scattering of an incident monochromatic plane wave projected upon N homogeneous, isotropic spheres of radius a, the volume-scattering function is given as

$$\beta(\Theta) = \frac{N}{2K^2} [i_1(\alpha, m, \Theta) + i_2(\alpha, m, \Theta)] . \qquad (3a)$$

In this representation, the volume-scattering function $\beta(\Theta)$ is the ratio of the flux scattered by a unit volume containing N particles of radius a into solid angle Ω in the direction Θ, to the flux incident on the unit cross section of this volume. Following the notation used by *Van de Hulst* [1957] and *Kerker* [1969], the remaining terms are defined as:

m = relative refractive index
λ = wavelength of incident radiation in the surrounding
   medium
$k = \dfrac{2\pi}{\lambda}$ = propagation constant in the medium

$\alpha = \dfrac{2\pi a}{\lambda} = ka$ = size parameter

Θ = scattering angle (the angle between the direction of
   propagation of the incident wave and scattered wave)
i_1 and i_2 are the intensity functions of the Mie scattering theory [*Van de Hulst*, 1957; *Kerker*, 1969].

If there is a range of particle sizes such that $N(a)da$ is the number of particles per unit volume with radii between a and $a + da$, then the volume-scattering function is given as

$$\beta(\Theta,\lambda,m) = \frac{1}{2k^2} \int_0^\infty N(a) [i_1(a,\lambda,m,\Theta) + i_2(a,\lambda,m,\Theta)] \, da . \qquad (3b)$$

If, in addition, there is a distribution of refractive indices for the suspended particles, then $\beta(\Theta)$ will be determined by an additional integration over this distribution. This additional complexity will not be considered in the following.
   The volume-scattering function, for a fixed λ and m, is determined experimentally by use of the operational definition (1). It may be determined theoretically, using known (or assumed) characteristics of the scattering particles, by use of Mie scattering theory (3). If trial characteristics of the suspended particulate material, used to theoretically calculate $\beta(\Theta)$, can be found which "fit" the experimentally measured volume-scattering function, then it is inferred that the particles have these trial characteristics. It is not yet clear if this "inverse problem" of determining properties of polydispersed particles from optical measurements can be uniquely solved, although useful information

on particulate distributions in the oceans have already been
made by workers [*Kullenberg*, 1970; *Kullenberg and Olson*, 1972;
Gordon and Brown, 1972; *Mertens et al.*, 1972] who have addressed
this problem.

It is clear, however, that the volume-scattering function
is a fundamental parameter in any effort to solve the inverse
problem. The experimental accuracy, precision, and completeness
with which $\beta(\theta)$ can be measured *in situ* will be important in
determining the completeness to which the inverse problem can be
solved and the accuracy of the extracted information.

The concentration and size distribution are two characteristics
of suspended particles which are of major importance. The total
number of suspended particles per unit volume of the medium, N_0,
is given as

$$N_0 = \int_0^\infty N(a)\,da \ . \tag{4}$$

Following the notation of *Kerker* [1969], the efficiency for
scattering, from a spherical particle of radius a, is

$$Q = \frac{1}{k^2 a^2} \int_0^\pi (i_1 + i_2) \sin \theta \, d\theta \ . \tag{5}$$

If there are N particles per unit volume of radius a (monodisperse
particle distribution) then the total volume-scattering function
(2) becomes

$$b = N\pi a^2 Q \quad . \tag{6a}$$

However, for a polydisperse particle distribution which is of
practical oceanographic interest, the total volume-scattering
function becomes

$$b = \int_0^\infty N(a)\pi a^2 Q \, da \quad . \tag{6b}$$

The behavior of Q, as a function of the parameter $\rho = 2\alpha(m - 1)$,
is discussed in detail by *Van de Hulst* [1947] and *Kerker* [1969].
Depending upon ρ, the value of Q reaches a maximum value and then
undergoes a damped oscillation about the limiting value of $Q = 2$.
If variations in Q with ρ are negligible (or neglected), then
(6a) and (6b) show that the total volume-scattering function b
is proportional to the total projected particle area per unit
volume of the medium. This relationship between b and the
effective particle area per unit volume is used to infer total
particle concentrations in the ocean [*Jerlov*, 1968, and references therein; *Carder et al.*, 1971].

Chandrasekhar [1950] introduces a phase function $p(\theta)$,
defined as

$$p(\theta) = \frac{4\pi \ \beta(\theta)}{b} \quad , \tag{7}$$

which is normalized so that

$$\int_{4\pi} p(\theta) \frac{d\Omega}{4\pi} = 1. \tag{8}$$

The phase function gives the rate at which energy is being scattered into an element of solid angle $d\Omega$ and in a direction θ. It is independent of N_0, the total concentration of particles per unit volume [*Deirmendjian*, 1963]. The shape of $p(\theta)$ versus θ is dependent upon the shape of the particle-size distribution and the Mie intensity functions. Using $(3b)$ and $(6b)$, the phase function is given as

$$p(\theta) = \frac{\frac{2\pi}{k^2} \int_0^\infty N(a)[i_1(a, \lambda, m, \theta) + i_2(a, \lambda, m, \theta)]\, da}{\int_0^\infty N(a)\, \pi a^2\, Q\, da}, \tag{9}$$

$$= \frac{1}{b} \frac{2\pi}{k^2} \int_0^\infty N(a)\, (i_1 + i_2)\, da \quad .$$

Comparison of experimentally determined phase functions with Mie theory calculations, using (9), can be used to infer information with respect to the particle-size distribution. The usual technique [*Kullenberg and Olsen*, 1972; *Gordon and Brown*, 1972] is to choose a convenient yet realistic analytic form to describe the particle-size distribution. Parameters of the chosen analytic size distribution are then varied until the calculated and measured values "fit".

SCATTERING VOLUME

For both the operational (1) and Mie scattering, $(3a)$ and $(3b)$, definitions of $\beta(\theta)$ given above, the concept of a scattering volume is crucial. Scattering volume holds the key to both correct theoretical interpretation and accurate experimental measurements. In a practical instrument used to measure $\beta(\theta)$, the scattering volume of a sample will have a finite size. The volume must be large enough to contain a sufficient number of scattering particles to produce a statistically detectable flux of scattered energy. Also, for Mie scattering theory to be applicable, the volume must be large enough and the particles far enough apart so that they can be considered as independent scattering centers. Further, the volume of space must be large enough so that a statistically similar sample of all the particles of the larger medium it represents are included. On the other hand, the particle concentration must be small enough that only incoherent or independent scattering need be considered. Hence, the volume must be small enough so that multiple scattering does not occur and so that the absorption due to the suspending medium within the volume is negligible. Thus, the optical path length of the volume must be small. The fidelity of meeting the conditions imposed upon the scattering volume sets limits to the accuracy of experimental scattering measurements and the theoretical conclusions inferred from them.

GENERAL- AND NARROW-ANGLE SCATTER METERS

Two instruments have been used for obtaining the volume-scattering function *in situ*: a narrow-angle scattering meter for measuring $\beta(\Theta)$ at forward angles of 0.086°, 0.172°, and 0.344° (1.5, 2.95, and 5.90 milliradians) and a general-angle scattering meter for determining $\beta(\Theta)$ between the limits $\Theta = 10°$ in the forward direction and $\Theta = 170°$ in the backward direction. For both instruments, careful attention has been given to defining the size of the scattering volume of the instrument and the limits to instrument accuracy imposed by this volume and the scattering and absorbing properties of the water.

Petzold [1972] has given a detailed description of the instruments, described a comprehensive validation experiment to demonstrate the reliability of the data obtained from the instruments, and presented $\beta(\Theta)$ data for representative ocean and laboratory waters. The validation experiment and laboratory tests were used to check the absolute accuracy and internal consistency of the scattering instruments of the Scripps Institution of Oceanography Visibility Laboratory. As a result of both laboratory tests and extensive field work, it is estimated that under working field conditions the instruments will measure $\beta(\Theta)$ *in situ* to an absolute accuracy of $\pm 15\%$. Exact estimates of error depend upon the scattering and absorbing properties of the water being studied.

There are few absolute determinations of $\beta(\Theta)$ with which to compare the results of the Visibility Laboratory instruments. *Kullenberg* [1968] and *Kullenberg and Olsen* [1972] report absolute values of the volume-scattering function obtained in the Sargasso Sea in 1966. The Visibility Laboratory's general-angle scatter meter was used by Austin in the Sargasso Sea during the *Discoverer* Expedition in 1971 (*Tyler*, 1973). When correction is made for the wave length-dependent scattering due to pure water [*Le Grand*, 1939], the volume-scattering functions as measured by Kullenberg and by Austin agree within the stated accuracy of the respective instruments [*Kullenberg*, personal communication]. Since these instruments were constructed and calibrated independently, this lends confidence to the results of both absolutely calibrated instruments.

SCATTERING-PHASE FUNCTION DATA AND
OPTIMUM SINGLE-ANGLE SCATTER METERS

Absolute *in situ* measurements of the volume-scattering function have required considerable care and skill on the part of the investigator and have been time-consuming to obtain. Because of this, investigators are frequently willing to forego information on par-

ticle-size distributions and to work, instead, to obtain an optical measure of the total particle concentration. As outlined a-
bove (6a) and (6b), the total volume-scattering function b may be
considered to be proportional to an effective projected area concentration of the particles. Following a suggestion by *Jerlov,*
[1953], a number of workers [*Tyler,* 1961; *Kullenberg,* 1968; *Carder
et al.,* 1971] have shown that the volume-scattering function measured at a single angle shows a linear correlation with the total
scattering function b. Based on these results a number of investigators are utilizing single-angle scattering meters as a measure of
the relative concentration of particle concentration in the oceans.
     It has, of course, been recognized that single-angle scattering
meters used for this purpose have limitations. First, in (6b), b
is not strictly proportional to particle concentration unless the
relative particle-size distribution remains the same as the concentration changes. *Owen* [1972], *Bader* [1970], and *Carder et al.,*
[1971] have presented data to support the hypothesis of a "quasi-
stationary distribution" of particle sizes in the upper layers of
the ocean. The extent to which single-angle scattering meters have
been successful is evidence that the hypothesis of a "quasi-stationary distribution" of particles is more or less true. Further, it
has been assumed that $\beta(\theta)$ at a fixed angle is directly proportional to b. Early data [*Jerlov,* 1953; *Tyler,* 1961] indicated that this
assumption holds well at a 45° angle of measurement. Support for
this choice of angle was given by *Deirmendjian* [1963], who plotted
phase functions for polydispersed systems of particles and found a
proportion between b and $\beta(40°)$. However, the proportion between
reported values of $\dfrac{\beta(45°)}{b}$ has a two to threefold variation and not

all workers agree that 45° is the optimum angle of measurement as
witnessed by the variety of instruments in use. A further serious
limitation of the single-angle instruments in use today is that the
results of few, if any, of these instruments can be directly compared. The lack of absolute calibration and variations in the
choice of fixed angle make direct comparison of results difficult.
     The volume-scattering function, the probability function, and
the phase function for three ocean water masses are shown in Figures 1, 2, and 3, respectively. The curves in Figure 2 show graphically the probability function as the ratio,

$$P(\theta) = \frac{2\pi \int_0^\Theta \beta(\theta) \sin\theta \, d\theta}{2\pi \int_0^\pi \beta(\theta) \sin\theta \, d\theta} = \frac{1}{b} \int_0^\Theta \beta(\theta) \sin\theta \, d\theta.$$

The integral, $2\pi \int_0^\Theta \beta(\theta) \sin\theta \, d\theta$, is that portion of the total

scattering coefficient which lies between zero and the angle θ.
The three water types span a wide range of optical properties. The
data for the Tongue of the Ocean is representative of deep, clear
ocean water; the data for the California coastal waters is an ex-

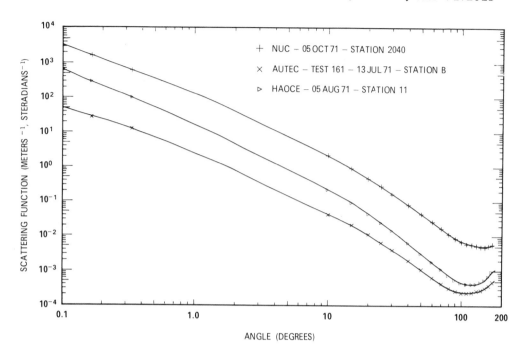

Fig. 1. Volume-scattering function β(Θ) (meters⁻¹ steradians⁻¹) versus scattering angle Θ (degrees) for three optically different water masses. AUTEC (x) = data from deep, clear ocean water of Tongue of the Ocean, Bahama Islands (24°29'N, 77°33'W); HAOCE (▷) = data from moderately productive Southern California coastal water (33°30.0'N, 118°23'W); NUC (+) = data from turbid San Diego Harbor water (32°42'N, 117°14'W). The symbols (x,▷,+) are data points obtained from the two scattering instruments at the Visibility Laboratory. The solid curves are computer plots of β(Θ) calculated from the raw data [*Petzold*, 1972].

ample of moderately productive surface waters; and the data for San Diego Harbor is an example of relatively turbid water. The values of the total attenuation coefficient c (measured with a beam transmissometer), the total scattering coefficient b (obtained from the data of Figure 1 using (2)) and the absorption coefficient a (obtained from the relationship $c = a + b$) are given in Table 1.

The choice of β(45°) as being most nearly proportional to b was based on early scattering data which did not include narrow-angle forward scattering. The complete data obtained with the Visibility Laboratory instruments (Figures 1 and 2) indicates that half of the scattering which contributes to the total scattering coefficient b originates forward of about 2° in coastal waters and forward of about 6° in clear ocean waters. Failure to account for this forward scattered energy when measuring b can account for large and systematic errors.

Table 1 and Figure 3 show that, whereas b varies by a factor of

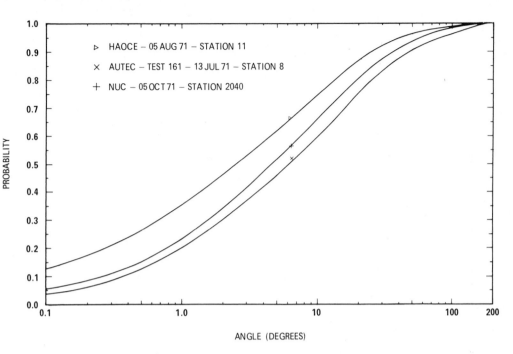

PROBABILITY

ANGLE (DEGREES)

Fig. 2. Probability function versus scattering angle calculated from the values of β(θ) given in Figure 1. AUTEC (x) = data from deep, clear water of Tongue of the Ocean, Bahama Islands (24°29'N, 77°33'W); HAOCE (▷) = data from moderately productive Southern California coastal water (33°30.0'N, 118°23'W); NUC (+) = data from turbid San Diego Harbor water (32°42'N, 117°14'W).

TABLE 1. Values of the Total Attenuation Coefficient c, the Total Scattering Coefficient b and the Absorption Coefficient a for the Data Shown in Figures 1, 2 and 3.

	$c[m^{-1}]$	$b[m^{-1}]$	$a[m^{-1}]$	b/c	a/c
AUTEC	.151	.037	.114	.247	.753
HAOCE	.398	.291	.179	.551	.449
NUC	2.190	1.824	.366	.833	.167

AUTEC = Data for Tongue of the Ocean, Bahama Islands (24°29'N, 77°33'W)
HAOCE = Data for offshore southern California (33°30.o'N, 118°23'W)
NUC = Data taken in San Diego Harbor (32°42'N, 117°14'W)

about 50 from clear to turbid water, the phase function varies by a factor of 5 or less for all angles. In particular, it should be noted that $p(45°)$ varies by more than a factor of 2, while the

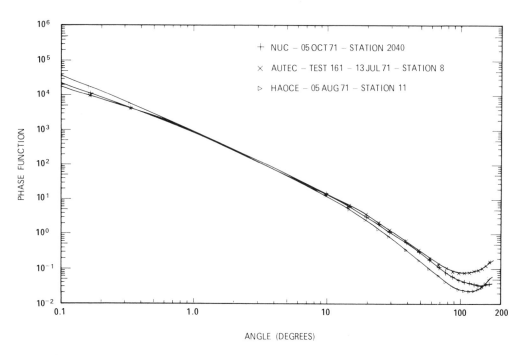

Fig. 3. Scattering-phase function $p(\Theta)$ versus scattering angle calculated from the values of $\beta(\Theta)$ given in Figure 1.

phase function in the 1° to 2.5° region varies by less than ±10%. There is an apparent constancy of $p(\Theta)$ for Θ = 1° - 2.5°, which is relatively independent of the particle-size distribution found over a wide range of types of ocean water. This indicates, for the quasi-stationary distributions of particles found in the upper layers of the ocean, that $\beta(\Theta)$ measured between 1° to 2.5° gives a more accurate (±10%) estimate of b than if measured at other angles (±100% or more for Θ = 45°). Thus, $\beta(1° - 2.5°)$ is also the optimum measurement for determining the effective projected area concentration of suspended particulate material.

These observations apply to data obtained in the upper layer of the oceans where the waters are relatively well-mixed and biological activity is maximum. As long as a "quasi-stationary distribution" of particle sizes is maintained within the water column, the above conclusion can be expected to remain valid. However, if the shape of particle-size distribution (or effective refractive index) changes significantly, a change in the shape of the phase function (9) would be expected. Thus, if there are significant changes in the shape of the particle-size distribution (as distinct from the absolute concentration of particles) within the water column, there will be no single angle for which the volume-scattering function will remain accurately proportional to b. There are no accurate and complete *in situ* determinations of $\beta(\Theta)$ below the upper few hundred meters of the ocean. Thus, it should be understood that a choice of an optimum angle for use in a single-angle scattering meter is made for these limited conditions.

CONCLUSIONS

The optical signature of the suspended material in a water mass has not yet been fully exploited. To fully extract all the information characteristic of particles in the ocean would require the following. First, it is necessary to have instrumentation for accurately and precisely measuring the total attenuation coefficient, total scattering coefficient, and the volume-scattering function (including narrow-angle forward scattering). Second, since solution of the inverse problem is largely dependent upon appropriate comparison with Mie scattering theory, a computer facility for the computation of the Mie scattering functions is required. Third, there is a need for laboratory studies of prepared polydispersed suspension of particles in sea water and their optical signatures. To achieve an optimization of information by optical methods one must compare data obtained in the field and in the laboratory using absolutely calibrated instruments, with computer simulated theoretical models of particle characteristics.

For information on only the effective projected area concentration of the suspended particles, a single-angle scattering instrument can be used. This instrument should be absolutedly calibrated so that results of various investigators can be compared and so that any large changes from the quasi-stationary distribution of particles can be detected. As noted above, an optimum angle for a single-angle instrument is between 1° and 2.5° for investigating the size distribution met in the upper layers of the ocean. Given an absolutely calibrated instrument and the usual practice of calibrating the instrument by comparison of the optical signal with gravimetric determination of particle weight, it may be possible to detect a gross change in size distribution and/or effective index of refraction. It is impossible to distinguish between a change in the concentration by using the effective projected area and a change in the shape of the size distribution from a single-angle instrument that measures only relative values.

REFERENCES

Bader, H., The Hyperbolic Distribution of Particle Sizes, *J. Geophys. Res.*, *75*, 2822, 1970.

Beardsley, G. F., H. Pak, K. L. Carder, and B. Lundgren, Light Scattering and Suspended Particles in the Eastern Equatorial Pacific Ocean, *J. Geophys. Res.*, *75*, 2837, 1970.

Carder, K. L., G. F. Beardsley, and H. Pak, Particle Size Distributions in the Eastern Equatorial Pacific, *J. Geophys. Res.*, *76*, 5070, 1971.

Chandrasekhar, S., *Radiative Transfer*, Dover Pub., Inc., New York, 393 pp., 1950.

Deirmendjian, D., Scattering and Polarization Properties of Poly-
    dispersed Suspensions with Partial Absorption, in *Electromagnetic
    Scattering*, edited by M. Kerker, 592 pp., Pergamon Press, New
    York, 1963.
Gordon, H. R., and O. B. Brown, A Theoretical Model of Light Scat-
    tered by Sargasso Sea Particulates, *Limnol. and Oceanogr.*, *17*,
    826, 1972.
Jerlov, N. G., Particle distribution in the ocean, in *Rep. Swedish
    Deep Sea Exped.*, *3*, 73-97, 1953.
Jerlov, N. G., *Optical Oceanography*, Elsevier, London, 194 pp.,
    1968.
Kerker, M., *The Scattering of Light*, Academic Press, New York,
    666 pp., 1969.
Kullenberg, G., Scattering of Light by Sargasso Sea Water, *Deep Sea
    Res.*, *15*, 423, 1968.
Kullenberg, G., A Comparison Between Observed and Computed Light
    Scattering Functions, *Institute of Physical Oceanography*, *Univer-
    sity of Copenhagen, Report No. 13*, 1970.
Kullenberg, G., and N. B. Olsen, A Comparison Between Observed and
    Computed Light Scattering Functions, II, *Institute of Physical
    Oceanography, University of Copenhagen, Report No. 19*, 1972.
Le Grand, Y., La pénétration de la lumière dans la mer, *Ann. Inst.
    Océanog.*, *19*, 393, 1939.
Mertens, L. E., D. L. Phillips, H. Gordon, O. Brown, and H. Bader,
    Measurements of the Volume Scattering Function of Sea Water,
    Technical Report 334, Project Deep Look, Range Measurement Labo-
    ratory, Patrick Air Force Base, Florida, 7 March 1972.
Owen, R. W., The Scattering of Light by Particulate Substances in
    the Sea, Ph.D. thesis, University of California, San Diego, 1972.
Petzold, T. J., Volume Scattering Functions for Selected Ocean Wa-
    ters, *Scripps Institution of Oceanography Ref. 72-78*, University
    of California, San Diego, 1972.
Tyler, J. E., Measurement of the Scattering Properties of Hydrosols,
    J. Opt. Soc. Amer., *51*, 1289, 1961.
Tyler, J. E., Data Report. Scientific Committee on Ocean Research.
    Discoverer Expedition May 1970, vol. 1, *Scripps Institution of
    Oceanography, Ref. 73-16*, University of California, San Diego,
    1973.
Tyler, J. E., and R. W. Preisendorfer, Transmission of Energy With-
    in the Sea, in *The Sea*, vol. 1, edited by M. N. Hill, Interscience
    Publishers, New York, 864 pp., 1962.
Van de Hulst, H. C., *Light Scattering by Small Particles*, John
    Wiley and Sons, New York, 470 pp., 1957.

MIE-THEORY MODELS OF LIGHT SCATTERING BY OCEAN PARTICULATES

HOWARD R. GORDON

University of Miami

ABSTRACT

The general application of Mie theory to the study of the volume-scattering function, VSF, for oceanic particles is discussed in detail. The problem examined is the combination of Coulter counter measurements and the theory in order to enhance our understanding of the size-refractive index distributions of the sea water particles. First single-component models (one scattering species) are examined and compared with the Sargasso Sea VSF. These models yield an average refractive index $\bar{m}$ relative to water. Questions concerning the meaning of $\bar{m}$ require the investigation of simple two-component models consisting of low-index organic particles, and high-index inorganics. These models are systematically studied and result (again for the Sargasso Sea) in the conclusion that the inorganic particles occupy large size ranges while the organic particles are confined to small sizes. Similar studies in the Tongue of the Ocean in the Bahama Islands indicate the necessity of three-component models, which place organic particles in small and large size ranges and inorganic particles in midsizes. This model predicts a VSF in good agreement with observation, and indicates that most of the scattering is due to the inorganic particles, with the large-size organic particles (phytoplankton) contributing only at small angles. This suggests the possibility of monitoring phytoplankton populations in coastal areas through simultaneous measurement of small- and large-angle scattering, for which a preliminary experimental verification is presented. A relaxation of the constraints on the three-component model consistent with the observed VSF has been effected and provides very general limitations on the size distribution in size ranges where it cannot be measured. A simple method for determination of the average refractive index of particles is presented and its limitations are discussed in detail.

73

INTRODUCTION

One problem of light-scattering research on ocean particles centers on finding more fundamental properties of the particles than the mere scattering data itself.

Ideally, the volume, shape, and composition of each scattering particle should be determined; however, there appears to be little hope of accomplishing this. The composition relates to the scattering process through the particle's refractive index m. It is assumed that the individual particles are uniform in the sense that they can be characterized by a single index of refraction. In principle, given m, the volume, and the shape, the volume-scattering function $\beta(\Theta)$ could be calculated but in practice this can only be done for spherical particles. (For a definition of $\beta(\Theta)$ and related quantities, see the paper by *R. C. Smith*, this volume.) It will be assumed here that the particles are spherical, in which case the volume gives the diameter. It is desired then to determine the size distribution of each species m of particles in the medium from the scattering data. Unfortunately, it is easy to show that even if all of the unknown indices are the same (i.e. if all particles have the same unknown m), it is impossible to uniquely determine the size distribution from $\beta(\Theta)$. Some knowledge of the size-refractive index distributions is required to invert the scattering data. Fortunately, the size distribution is available in the range 0.4 to 400μ from the Coulter counter [*Bader*, 1970] which sizes particles according to volume, independent of composition. An example of size distribution using a Coulter counter on particles from the Tongue of the Ocean, Bahama Islands, is given in Figure 1. It should be noted that since there is very little information available in sizes below 0.4μ, it is hoped that a theory of light scattering will, as a byproduct, place certain limits on the distribution in this size range.

SINGLE-COMPONENT MODEL

The problem of explaining light-scattering data is attacked by combining total size distributions from the Coulter counter with Mie scattering theory [*Mie*, 1908] to find the refractive index distribution of the particles. As a first approximation, it is reasonable to assume that all of the particles have the same refractive index (composition) and to determine the value of m required to fit experimental data. One might call m the average or "significant" [*Zaneveld and Pak*, 1973] refractive index. For this purpose, *Kullenberg's* [1968] volume-scattering functions for the Sargasso Sea and size distributions for the same area (measured in 1970) were used. The observed size distribution was

$$N_{>D} = KD^{-c} \quad ,$$

where $N_{>D}$ is the number of particles per mℓ larger than a given diameter D (microns) and c and K are constants, with $2.4 \underline{<} c < 3.1$ and

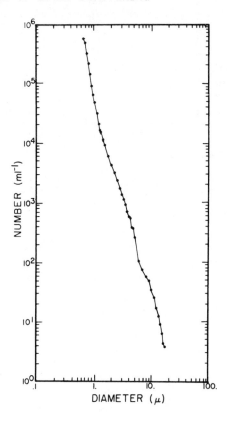

Fig. 1. An example of a Coulter counter size-distribution from the surface at High Cay, Tongue of the Ocean, Bahama Is., 0800 2 June 1971. $N_{>D}$ is plotted against diameter.

$8 \times 10^3 < K < 20 \times 10^3$ particles/mℓ. This distribution was found to maintain 0.8μ which, at that time, was the lower size limit of the Coulter counter. Using distributions of this form, $\beta(\theta)$ can be calculated in terms of c, K, m, and the lower limit of the size distribution D_1 (which is unknown). This calculation has been carried out for several indices: 1.01, 1.01-.01i, 1.01-.1i, 1.03, 1.03-.01i, 1.05, 1.05-.01i, 1.05-.1i, 1.10, 1.15, 1.20 and 1.25 where $i = \sqrt{-1}$; for $2 \leq c \leq 4$, $D_1 = 0.08$, 0.09, 0.1, 0.5, 1, and 5μ, at wave lengths of 425, 530, and 625nm. The results of these calculations [*Gordon and Brown*, 1972] show that $m = 1.05-.01i$ is the only index which can reproduce $\beta(\theta)$ for the range of K and c values observed. The calculated distributions were

$$N_{>D} = \frac{11 \times 10^3}{D^3} \quad m\ell^{-1} \quad 0.08\mu < D < 10\mu$$

and the observed and calculated volume-scattering functions are given in Table 1. The low value of the index implies the existence of a very large organic component in the suspension, since the index of minerals is generally equal to or larger than 1.15. Since both organic particles and inorganic particles may contribute strongly to the scattering, a two-component model was developed to try to separate effects of these constituents.

TABLE 1. Observed and Calculated Volume-Scattering Functions for the Single-Component Model

Quantity*	Observed	Calculated
$\beta(1°)$	11,000	11,264
$\beta(10°)$	380**	392.7
$\beta(30°)$	22	17.5
$\beta(45°)$	10.3	4.82
$\beta(90°)$	0.5	0.502
b_s	0.023	0.023
b_a	-	0.020

* $\beta(\theta)$ in units of $10^{-4}m^{-1}$, ster, b_s and b_a in m^{-1}.
** Interpolated from *Kullenberg* [1968].

TWO-COMPONENT MODELS

The two-component models divide the scattering particles into two species: organic particles and inorganic particles, or minerals. It is assumed that organic particles have an index (m_0) 1.01-.01i, i.e., very near that of water with some absorption. The index of the inorganic particles (m_I) is taken to be 1.15 (quartz relative to water). The total size distribution

$$N_{>D} = \frac{K}{D^3} , \quad 0.1 \leq D \leq 10\mu$$

was chosen to roughly conform to the previous model. The following three cases were considered.

(1) The total size distribution given above is divided into a fraction with diameters smaller than D' (henceforth called the small fraction) composed of low-index particles m_0, and a fraction with diameters larger than D' (large fraction) composed of high index material m_I.

(2) The second total size distribution is similar to that of (1) above; however, the m_I particles are in small fraction, and the large fraction is comprised of m_0 particles.

(3) The third total size distribution (called uniform) has

both species of particles distributed throughout the entire size range (0.1 to 10μ), but differing in concentration. Letting K_0 and K_I be the number of particles larger than 1μ of the low- and high-index particles, respectively, for this case

$$N_{j>D} = K_j/D^3 \qquad\qquad 0.1 \leq D \leq 10.0μ$$

where the index j = 0 or I. The Coulter counter measures particles independent of their composition. It therefore yields the same distribution for the three cases, since it is required, in case (3), that $K_0 + K_I = K$.

These distributions were combined with the Mie theory to find $β(θ)$ for volume concentrations of the organic particles relative to the total (v_0/V) from 0 to 1 in increments of 0.1. For cases (1) and (2), v_0/V is a function of D' only while for case (3), $v_0/V = K_0/K$. Note that for cases (1) and (2), the free parameters in the model are D' and K, while for case (3) the parameters are K_0 and K_I. The resulting volume-scattering functions for the three cases are given in *Brown and Gordon* [1973]. The comparison with Kullenberg's $β(θ)$ is shown in Figure 2, where it should be noted that $β(θ)$ is normalized to $β(1°)$. It is clear that case (1) with 0.7 < v_0/V < 0.8 (B,C), or case (3) with 0.9 < v_0/V < 1.0(D,E) provide volume-scattering functions of the correct shape. Curve A with v_0/V = 0.9 fits the data poorly, but is included in the figure since it gives the best fit for case (2). The range of v_0/V for these two cases provides the range of D' and K_0/K. To find which model is best, the calculations are denormalized using $β(1°)$ for the three cases given in Figure 3. From this, it is found that case (1) requires a K of 1.61 x 10^4 $mℓ^{-1}$ while for case (3), a K of 4.78 x 10^4 $mℓ^{-1}$ is appropriate. Hence only case (1) will fit the observed Coulter-counter data. The resulting distribution is then

$$N_{>D} = \frac{16.1 \times 10^3 mℓ^{-1}}{D^3} \qquad 0.1 \leq D' \leq 1.0μ$$

$$m = 1.01 - .01i \qquad\qquad 0.1 \leq D \leq D'$$
$$m = 1.15 \qquad\qquad D' \leq D \leq 10μ$$

where v_0/V = 0.7 gives D' = 2.5μ. The observed and calculated scattering fractions and the contribution from the high-index particles (1.15) are given in Table 2. It is interesting that over 90% of the scattering is due to the minerals which account for only 30% of the particulate volume. Due to the small contribution of the m_0 particles, it is clear that these could occur in sizes larger than 2.5μ without significantly changing $β(θ)$ while, on the other hand, large quantities of minerals in sizes smaller than 2.5μ must be ruled out. In fact, if one assumes that minerals exist in sizes smaller than D' and distributes them according to

$$N_{>D} \sim \frac{1}{D^\alpha} \qquad ,$$

we find that as long as $\alpha \leq 1/2$, the agreement between the observed and calculated $\beta(\theta)$ is preserved.

The results from these two-component models depend, of course,

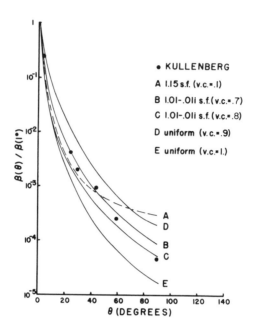

Fig. 2.   Comparison of the shapes of $\beta(\theta)$ for the various two-component models with experimental data for the Sargasso Sea from *Kullenberg* [1968].   V.C. = the volume concentration of the small fraction.

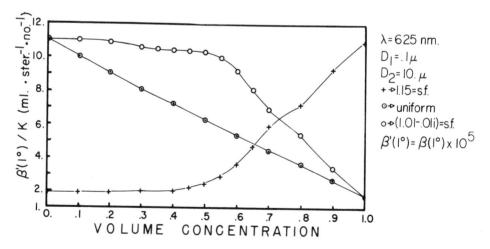

Fig. 3.   Variation in $\beta(1°)/K$ with volume concentration of the small fraction for the three two-component models considered.

TABLE 2.   Observed and Calculated Volume-Scattering Functions
            for the Two-Component Model.

Quantity*	Observed	Calculated	% from m = 1.15
$\beta(1°)$	11,000	11,000	99.
$\beta(10°)$	380**	352	90
$\beta(30°)$	22	34.7	95
$\beta(45°)$	10.3	7.85	94
$\beta(60°)$	2.8	2.53	92
$\beta(90°)$	0.5	0.978	95
b_s	0.023	0.0268	95
b_a	-	0.0207	0

* $\beta(\theta)$ in units of $10^{-4}m^{-1}\,ster^{-1}$ and b_s, b_a in m^{-1}.
** Interpolated from *Kullenberg* [1968].

on the value of the index assigned to the organic component.   If
m_0 is decreased, the results will be unchanged while increasing m_0
will cause D' to increase; in fact, if m_0 is increased to 1.05-
.01i, D' will increase to 10μ and the result is the single-component model.   Varying the high index in the range 1.15 to 1.20 will
not significantly change the results [*Gordon and Brown*, 1971].   Reducing the lower limit (0.1μ) and increasing the upper limit (10.0μ)
will also have little influence on the conclusions made using the
model.
     In the absence of simultaneous Coulter-counter and scattering
data for the Sargasso Sea, it is useless to try to construct more
complex models.   However, in the Tongue of the Ocean, Bahama Is.,
such data is available and is presented below.

                    THREE-COMPONENT MODELS

     During 1971, a series of Coulter-counter and light-scattering
measurements were carried out in Tongue of the Ocean, Bahama Is.,
[*Gordon et al.*, 1972].   Figure 4 shows the average size distributions over several seasons for Tongue of the Ocean.   The segmented
nature of these curves is to be noted.   On one occasion various
elements of the scattering matrix were measured at 488 nm by *Mertens and Phillips* [1972] simultaneously with the size distribution
at High Cay.   In trying to fit the size distribution and the scattering functions, it was found that single-component models were not
appropriate.   Since the distributions show three distinct sections
(three values of c), it was decided to attempt a three-component
model by assigning different indices to each section of the distribution.   The model consisted of using actual Coulter-counter
measurements with the refractive indices of the particles distributed according to:

$$m = 1.01 \qquad\qquad 0.65 \leq D \leq 1.25\mu$$
$$m = 1.15 \qquad\qquad 1.25 < D < 3.75\mu$$
$$m = 1.01 - .01i \qquad\quad 3.75 \leq D \leq 17\mu$$

The lower-limit 0.65μ was the smallest size that could be measured on the counter and the upper limit was the largest size actually measured. The sizes at which the indices change (1.25 and 3.75μ) were chosen to correspond to diameters at which the slope-changes are observed in the data (Figure 4). The choice of indices was based on the following observations:

1) The very small particles occur in such large numbers that they must have a low index of refraction; otherwise, the water would be nearly opaque. If this index were as large as 1.05, the numbers would completely dominate the scattering and would result in an incorrect shape for $\beta(\theta)$.

(2) The small particles are probably organic in nature, since they do not appear to settle appreciably over time scales of months and if filtered out of a sample, they return in about a day.

(3) If there are phytoplankton in a sample, they will be distributed in large sizes and the refractive index for phytoplankton has been shown (on the basis of small-angle scattering) to be in the range 1.01 to 1.03 [McCluney, 1973]. Because phytoplankton absorb some light, a small amount of absorption is added to the index.

(4) The minerals are placed in midsizes consistent with the two-component models.

The elements M_{11} and M_{22} of the Stokes' scattering matrix computed for this model are presented in Figure 5 for each component of the model and for pure water.

In general, $\beta(\theta) = M_{11}$, while, for spherical particles, $M_{11} = M_{22}$, so $\beta(\theta)$ is M_{11} or M_{22}. Comparison with the observed matrix elements is given in Figure 6. The model clearly indicates that the minerals dominate the particle scattering at all angles larger than about five degrees, that the small-size organic particles contribute essentially nothing to $\beta(\theta)$ and that the phytoplankton (large organic particles) can contribute only at very small angles.

The results of this model suggest the possibility of monitoring phytoplankton populations through the simultaneous observation of scattering at both small and large angles. McCluney [1973] compared scattering at 2°, 45°, and 90° with phytoplankton microscope counts, n, and found that $\beta(2°)$ follows the variations in n far better than $\beta(45°)$ or $\beta(90°)$. In this experiment, $\beta(90°)$ appeared to be proportional to $\beta(45°)$. This then may be considered a preliminary verification of the three-component model.

Even though this model fits the data well, it is unpleasant in that (1) there are no particles smaller than 0.65μ, (2) there are no minerals smaller than 1.25μ, and (3) there are no organic particles in the range $1.25-3.75\mu$. In order to try to rectify this, these particles were included subject to the constraint that the calculated *total scattering coefficient* must still agree with the experimental value, taking into account its uncertainty. The detailed process of accomplishing this has been described by Brown [1973] and results in limitations on the possible size distribu-

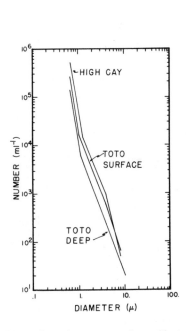

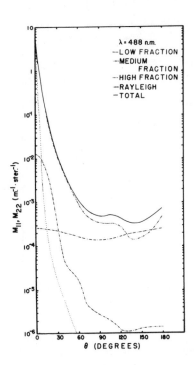

Fig. 4. Average size distributions for the Tongue of the Ocean, Bahama Is., (TOTO)

Fig. 5. Computed M_{11} and M_{22} elements of the scattering matrix of High Cay water (solid curve) with the contributions from the three components of the model, and Rayleigh scattering by the water.

tions. On this basis, a proposed size-refractive index distribution is presented in Figure 7.

The solid line in Figure 7 is the observed data, while the dashed lines refer to possible extensions of the distribution in sizes not measured. The arrow pointing to the left means these segments of the distribution can extend to zero diameter. Briefly, the calculations show that (1) the organic particles below 0.65μ could be distributed with $c\approx5$ to 6 down to 0.2μ or could have $c<3$ and be distributed from 0.65μ to zero size; that (2) the minerals below 1.25μ must be distributed with c<.5; that (3) the minerals above 1.25μ could be distributed with 2.7<C<3.5 depending on the amount of organic material present in this size range; that (4) the particles larger than 3.75μ, assumed to be phytoplankton, could be minerals or a mixture of minerals and phytoplankton, in which case the scattering at small angles would be larger due to the increased index. Since this size-refractive index distribution has been derived in a very general way and agrees with the experimental data, it is felt that it represents a close approximation to reality.

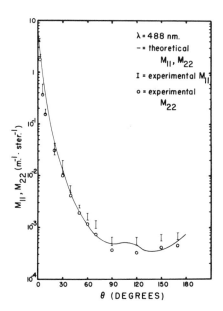

Fig. 6.   Comparison between
observed and computed values of
M_{11} and M_{22} for High Cay water.
Redundant determination of M_{11}
was possible from the data and
the vertical bars on M_{11} give
the smallest and largest values
of this quantity, providing an
estimate of the accuracy of
M_{11} and M_{22}.

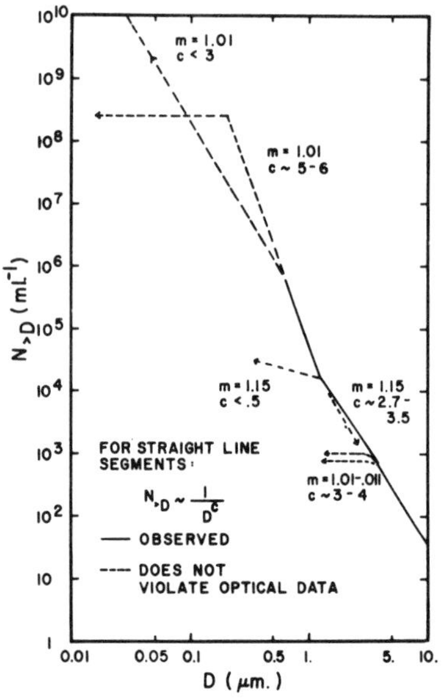

Fig. 7.   Proposed size-
refractive index distribution
on the basis of relaxation of
the constraints on the three-
component model.

EXPONENTIAL DISTRIBUTION FOR MINERALS

The model indicates that most of the scattering is due to
minerals and, considering the possible mineral distribution, it
is tempting to try to use the exponential distribution

$$N_{>D} = N_0 e^{-AD}$$

suggested by *Zaneveld and Pak* [1973] to analyze scattering
experiments in the ocean. When this is done, the resulting $\beta(\theta)$
agrees well with that using the three-component model and suggests
a method for determining an average refractive index for the
particles (excluding the organic particles <1μ) based on simple

measurements. The technique consists of measuring $\beta(45°)$, N_0, and
A, and comparing $\beta(45°)/N_0$ with calculated values of this quantity
for various refractive indices. Calculated values of $\beta(45°)/N_0$
as a function of A for several indices is given in Figure 8, along
with some experimental data from Tongue of the Ocean [*Gordon et al.*,
1972]. The Coulter-counter data from 7 samples was analyzed by
least squares for N_0 and A in three ways: (1) using all data in
sizes larger than 2μ; (2) using data only from 2≤D<3μ; and (3)
using data only from 2<D<4μ. All of these data are presented
in Figure 8, wherein it is clear that the resulting indices are
somewhat insensitive to the size range used to fit the data.

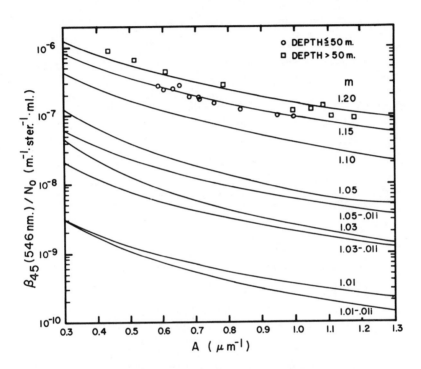

Fig. 8. $\beta(45°)$ at 546 nm divided by N_0 plotted against A for
the exponential distribution with various refractive indices m.
The data is from the Tongue of the Ocean, Bahama Is.

The results clearly favor an average index of refraction of
1.15 for surface water and of 1.17 to 1.22 for deep water. The
surface particle index is, of course, consistent with the three-
component model. The apparent increase with depth of the average
index can be explained by noting that near the surface there will
be larger quantities of organic particles (D>1-2μ) than at great
depth. These organic particles contribute to N_0 but not to $\beta(45°)$
causing $\beta(45°)/N_0$ to increase with depth for a given A. This then

implies that the average index determined in this manner is an indication of the relative abundance of minerals and organic material in the water in sizes larger than about 1μ. A similar procedure using $\beta(90°)/N_0$ does not seem to work well. In this case, the theoretical curves are not well separated and, in fact, intercept one another. Further, the experimental data does not seem to cluster near any particular refractive index. This could be due to the shape of the particles which can cause strong departures from spherical particle calculations at large scattering angles.

Although the procedure described appears to work well, it should be pointed out that it does not reproduce the wave-length variation of $\beta(45°)$. *Zaneveld and Pak* [1973] and *Gordon et al.* [1972] find $\beta(45°)$ at 546 nm. $\beta(45°)$ at 436 nm is usually about 1.25, while the above procedure with constant m indicates a value less than 1 for this ratio. If this discrepancy is explained by assuming m increases with wave length, it is found that a very large increase in m(from 1.15 to 1.19) is required.

It should be noted in Figure 8 that, for a constant total number of particles (N_0), $\beta(45°)$ is strongly dependent on A. For $m = 1.15$, a change in A from 0.9 to 0.7 can cause the scattering at 45° to double for a constant number of particles. Thus, it may be unreasonable to interpret changes in scattering directly as changes in particle concentration. This makes the interpretation of single-angle scattering data difficult. It may, however, be possible to rectify this by measuring scattering at two angles.

CONCLUSIONS

The results presented here demonstrate that scattering theory can be usefully applied to the study of oceanic-particle suspensions, as long as some knowledge of the size or refractive distribution is available. It is felt that the most significant results are that (1) the scattering is strongly dominated by the minerals; (2) the mineral-size distribution must roll over from a c of about 3 above 1μ to a c less than 1/2 below 1μ; and (3) phytoplankton concentration can be monitored by simultaneous observation of small- and large-angle scattering. For the routine analysis of scattering data, methods similar to that presented in Figure 8 seem superior.

Since minerals produce most of the scattering, one should be able to compute the mineral-size distribution from measurements of scattering at several angles, using a refractive index in the range 1.15 to 1.20. If an exponential form for the size distribution is assumed, then measurement of $\beta(\theta)$ at two angles would be necessary to determine the distribution parameters; however, in the light of the above discussion, measurement of scattering at large angles should be discouraged.

Several questions remain to be answered:

(1) What is the actual refractive index and the nature of the large number of apparently organic particulates < 1μ?

(2) Can the relative concentrations of minerals and organic

particles in sizes >1μ actually be determined using scattering theory?

(3) Is it realistic to choose an analytic form for the suspended particle-size distribution and how strongly do conclusions based on such a choice depend on the specifics of the distribution?

(4) Can adequate size distributions be determined by measuring $\beta(\Theta)$ at a few selected angles (such as 2) and which angles are best for such a determination?

It is felt that when these questions are answered, the precise interpretation of oceanic light scattering observations will be possible, making them more valuable in oceanography.

ACKNOWLEDGEMENTS

The author wishes to sincerely acknowledge his former students: Dr. Otis B. Brown, who was responsible for the Mie calculations and much of the analysis presented here, and Dr. W. Ross McCluney whose study of the optical properties of several marine phytoplankton yielded their Mie refractive indices. He also gratefully acknowledges the tireless work of Dr. Henri Bader in providing accurate Coulter-counter size distributions, and Dr. L. E. Mertens for allowing access to some of his data not included in *Mertens and Phillips* [1972]. The work received support from the Office of Naval Research, the Department of the Air Force, and the National Institutes of Health. This paper is contribution No. 1692 of the University of Miami, Rosenstiel School of Marine and Atmospheric Sciences.

REFERENCES

Bader, H., The hyperbolic distribution of particle sizes, *J. Geophys. Res., 75*, 2822, 1970.

Brown, O. B., Light scattering by ocean-borne particulates, PhD Thesis, University of Miami, 1973.

Brown, O. B. and H. R. Gordon, Two-component Mie scattering models of Sargasso Sea particulates, *Applied Optics, 12*, 2461, 1973.

Gordon, H. R. and O. B. Brown, Small-angle Mie scattering calculations for low-index hydrosols, *J. Opt. Soc. Am., 61*, 1549, 1971.

Gordon, H. R. and O. B. Brown, A theoretical model of light scattering by Sargasso Sea Particulates, *Limnol. Oceanogr., 17*, 826, 1972.

Gordon, H. R. and O. B. Brown and H. Bader, An experimental study of suspended particulate matter in the Tongue of the Ocean and its influence on underwater visibility, *Final Report* Contract No. F08605-7-C-0028, Air Force Eastern Test Range Measurements Laboratory, 1972.

Kullenberg, G., Scattering of light by Sargasso Sea Water, *Deep Sea Res., 15*, 423, 1968.

McCluney, W. R., Small-angle light scattering studies of marine phytoplankton, Ph.D Thesis, University of Miami, 1973.

Mertens, L. E. and D. L. Phillips, Measurements of the volume scattering function of sea water. *Tech. Rep.*, *334*, Range Measurements Laboratory Patrick AFB, Florida, 1972.

Mie, G., Beitrage zur optik truber Medien, speziell kolloidalen Metal-losingen, *Ann. Phys.*, *25*, 377, 1908.

Zaneveld, J. R. V., and H. Pak, Method for determination of the index of refraction of particles suspended in the ocean, *J. Opt. Soc. Am.*, *63*, 321, 1973.

Spatial Distribution of the Index of Refraction of Suspended Matter in the Ocean

J. RONALD V. ZANEVELD

Oregon State University

ABSTRACT

The use of the index of refraction of suspended particles as a tracer of water masses and for the analysis of the composition of particulate matter is discussed. A method for determining the index of refraction of suspended particles is presented. In order to determine an index of refraction for a sample, the ratio of light scattered at 45° for two wavelengths and the particle-size distribution must be known. By using a ratio, the difficult problem of determining absolute scattering values is avoided. In order to evaluate the index of refraction as an oceanic parameter, two widely different oceanic areas — the Oregon coastal area and the Equatorial Pacific near the Galapagos Islands — are examined. The spatial distribution of the index of refraction in these areas is studied in relation to ocean dynamics and to the nature of the suspended material. The limitations and advantages of using the index of refraction of suspended particles as an oceanic parameter are presented.

INTRODUCTION

The concentration and size distribution of particulate matter are important parameters in the ocean. These parameters largely determine the inherent optical properties in the ocean: light scattering and light attenuation. In order to operate any optical device in the undersea environment, the basic behavior of the inherent optical properties and hence the basic behavior of particulate matter must be understood. The concentration of particulate matter is a vital parameter in the study of sedimentation processes.

Optical methods provide quick survey tools for obtaining parameters related to the concentration of particulate matter. Coulter counters are used to obtain the size distribution of particles larger than a given size. No parameter is currently available which routinely and accurately gives information on the nature of the particulate matter. Of particular interest is the proportion of organic material in a particle sample. The nature of the particulate matter can be studied by filtering seawater samples. Implementation of these methods can be cumbersome, however.

The index of refraction of a particle is a parameter related to the nature of the particle. Particles of organic matter generally have low indices of refraction whereas inorganic material has a higher index of refraction. The index of refraction would be a good parameter in the study of particle dynamics if it could be routinely determined. This paper presents a study of the feasibility of using the index of refraction as a routine parameter in the study of particulate matter and its distribution in the ocean.

THE INDEX OF REFRACTION

The particulate matter in the oceans consists of many different materials ranging from live plankton to clay minerals. To properly describe the index-of-refraction characteristics of a particle sample, the index-of-refraction distribution for each size range in the particle-size distribution must be given. Our current research is directed toward this goal. At the present time, however, the best that can be done routinely is to obtain a representative index of refraction for an entire particle sample.

Care must be taken when the index-of-refraction properties of a sample are described by means of one number. Light scattering and attenuation properties do not average linearly over a collection of particles. Using different methods to obtain an "average" index of refraction for a sample of particles will thus lead to different average indices of refraction.

The earliest methods for determination of the index of refraction [*Sasaki et al.*, 1960; *Kullenberg*, 1970] use *Mie* [1908] theory to calculate volume-scattering functions for various indices of refraction when the size distribution is assumed to be known. The index of refraction which produced the volume-scattering function most closely resembling the observed volume-scattering function is chosen to be representative for the entire sample. This method is cumbersome, although it is reasonably accurate if complete information on the volume-scattering function and the particle-size distribution is available.

Zaneveld and Pak [in preparation, 1973] have developed a method which permits the calculation of an index of refraction for a particle sample based on the wave-length dependence of light scattering at 45°. *Brown and Gordon* [in preparation, 1973] have suggested a method using specific light scattering at 45° — that is, the volume-scattering function at 45° divided by the total number of particles present.

The method of Zaneveld and Pak may be less accurate for relative indices less than 1.03 or for absorbing particles due to the assumption that the total scattering coefficient is proportional to the light scattered at 45°. *Gordon et al.* [1971] have shown that large numbers of small particles that do not contribute greatly to light scattering do probably exist. Large numbers of small particles would drastically change the specific light scattering and the index of refraction.

Morel [1973] has made a thorough investigation of the scattering characteristics of sea water suspensoids using Mie-scattering calculations. Comparing experimental observations of light scattering with scattering derived from theory, Morel concludes that the average relative index of refraction of sea-water suspensoids usually lies in the range 1.02 to 1.05. This is exactly the range of indices most commonly observed using the method of Zaneveld and Pak.

A Method for the Determination of the Index of Refraction of a Particle Sample

The method *Zaneveld and Pak* [in preparation, 1973] used to obtain the indices of refraction in the study of the spatial distribution of this parameter will be briefly described here. From the exact Mie theory, *Van de Hulst* [1957] obtained approximate expressions for the scattering and attenuation efficiencies of spheres when the relative index of refraction is close to 1 and the particles are larger than the wave length of light. Integrating the efficiencies over a particle-size distribution gives the total attenuation and scattering efficiencies of the particle sample. In order to describe the particle-size distribution as a parameter, an exponential distribution was chosen, permitting writing of the cumulative particle-size distribution as

$$g(D) = Ne^{-AD} , \tag{1}$$

where $g(D)$ is the number of particles having diameters larger than D. N is the total number of particles and A is a parameter characterizing the shape of the size distribution. Using this size distribution and Van de Hulst's approximations for the attenuation and scattering efficiencies, the following expressions for the total attenuation coefficient c_p and the total absorption coefficient a_p of the particle-size distribution are obtained.

$$a_p = \frac{N\pi}{2} \left(\frac{1}{A^2} - \frac{1}{(A + 2k \tan \beta)^2} \right) \tag{2}$$

$$c_p = NA\pi \left[\frac{1}{A^2} - 2\cos^2\beta \frac{A + k\tan\beta}{[(A + k\tan\beta)^2 + k^2]^2} \right.$$

$$+ \frac{\sin 2\beta}{2k} \frac{[(A + k\tan\beta)^2 - k^2]}{[(A + k\tan\beta)^2 + k^2]^2}$$

$$- \left(\frac{\cos\beta}{k}\right)^2 \cos 2\beta \frac{A + k\tan\beta}{(A + k\tan\beta)^2 + k^2}$$

$$\left. - \cos^2\beta \frac{\sin 2\beta}{k} \frac{1}{(A + k\tan\beta)^2 + k^2} + \frac{\cos\beta}{k}^2 \frac{\cos 2\beta}{A} \right] \quad (3)$$

When n_p is the real part of the particle index of refraction, $n_p{}^1$ is the imaginary part of the index of refraction; m_w is the real index of refraction of water; λ_{vac} is the wave length of light in vacuum.

$$k = \frac{2\pi}{\lambda_{vac}} |n_p - m_w|$$

$$\tan\beta = \frac{n_p{}^1}{|n_p - m_w|}$$

It can be seen from (2) and (3) that if the particulate absorption and attenuation coefficients are known the real and imaginary parts of the index of refraction of the particles may be calculated if the particle-size distribution is known.

Unfortunately, the absorption coefficient and attenuation coefficient due to particulate matter alone are not easy to measure in the ocean at present, and approximations must be made.

The particle-absorption coefficient in the ocean is generally small; hence most scattering calculations in the oceans are carried out for non-absorbing spheres ($n_p{}^1 = 0$). Using this approximation, a_p becomes zero and the equation for the particle-attenuation coefficient, which now equals the total scattering coefficient b_p becomes

$$c_p = b_p = N\pi \left(\frac{1}{A^2} + \frac{k^2 - A^2}{(A^2 + k^2)^2} \right) \quad (4)$$

Equation (4) is plotted on Figure 1. It is seen that if the particle-size distribution characterized by A and N and the total scattering coefficient for particles are known, one may obtain a value for the index of refraction of the particulate matter.

The total scattering coefficient for particles is difficult to determine routinely. In order to use the above theory, an hypothesis first postulated by *Jerlov* [1953] is used. The hypothesis states that light scattering β_{45} is proportional to the total scattering coefficient b_p. Using this assumed relationship

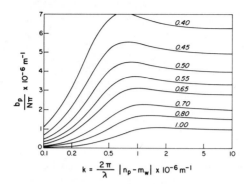

Fig. 1. The specific total scattering coefficient $b_p/N\pi \times 10^{-6}m^{-1}$ as a function of the index-of-refraction parameter $k = \frac{2\pi}{\lambda}\left|n_p - m_w\right| \times 10^{-6}m^{-1}$ and parameter A of the exponential particle-size distribution for the case of non-absorbing spherical particles $(np^1 = 0)$.

between β_{45} and b_p, we postulate:

$$\frac{b_p(\lambda_1)}{b_p(\lambda_2)} = \frac{\beta_{45}(\lambda_1)}{\beta_{45}(\lambda_2)} \tag{5}$$

Recent Mie-scattering calculations (Gordon, personal communication) have shown (5) to be correct within 10% except when the relative index of refraction is less than 1.03, in which case the Mie calculations themselves are somewhat in doubt, the largest particle diameter included being 20μm. Combining (4) and (5) for two wave lengths, λ_1 and λ_2, gives

$$\frac{b_p(\lambda_1)}{b_p(\lambda_2)} = \frac{\beta_{45}(\lambda_1)}{\beta_{45}(\lambda_2)} = \frac{1/A^2 + (k_1^2 - A^2)/(A^2 + k_1^2)^2}{1/A^2 + (k_2^2 - A^2)/(A^2 + k_2^2)^2} \tag{6}$$

Using this equation has several practical advantages. First, the constant of proportionality between β_{45} and b_p need not be known provided it is independent of wave length. Secondly, the ratio of $\beta_{45}(\lambda_1)/\beta_{45}(\lambda_2)$ does not depend on the absolute values of the light scattering, but on the relative intensities of light scattered at two wave lengths. A graphical solution of (6) is shown on Figure 2. Using this solution or a numerical approach one can readily obtain an index of refraction for a particle sample.

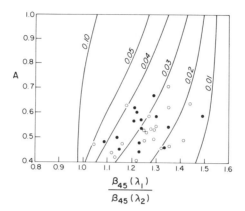

Fig. 2. Parameter A of the exponential particle-size distribution as a function of the ratio of light scattered at 45° for two wave lengths $\beta_{45}(\lambda_1)/\beta_{45}(\lambda_2)$, with the difference between the particulate and water indices-of-refraction $|n_p - m_w|$ as a parameter. Dots and circles represent samples taken off the coast of Ecuador.

Spatial Variation of the Index of Refraction

If the index of refraction is to be used as a means of identifying particulate matter as to possible origin, such a correlation must be shown to exist in the open ocean. High indices of refraction are expected to be related to high concentrations of inorganic material or materials of terrestrial origin. Preliminary results from an investigation off the mouth of the Columbia River on the Oregon Coast indicate that the index of refraction of suspended particles in the tongue of the Columbia River is generally higher than the index of purely oceanic particles.

The index of refraction was calculated for samples obtained during the Coastal Upwelling Experiment off the coast of Oregon. Typical distributions of total particle count, light scattering at 45°, and the index of refraction are shown in Figures 3, 4 and 5. Clearly distinguishable is the upwelled water, with a low concentration of particulate matter and, hence, low light scattering. This tongue is well correlated with a tongue of water in which the index of refraction (plotted is the difference between the particle and water indices of refraction) is considerably lower than the surrounding water. Surface waters (influenced by runoff) and near-bottom waters contain particulate matter having a higher index of refraction. Near-bottom water shows a higher concentration of particulate matter than water at intermediate depths. The near-bottom particulate matter has probably been stirred up from the bottom and thus may be expected to contain a larger concentration of inorganic material which is reflected in the index of refraction. The index-of-refraction distribution is thus not unreasonable in the light of expected physical processes.

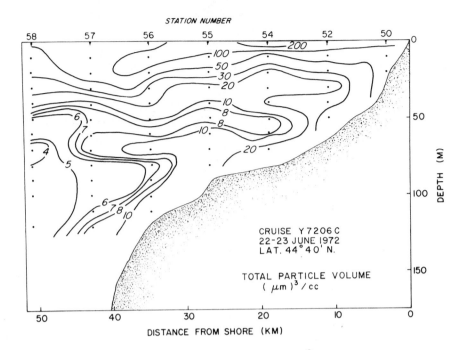

Fig. 3. Total particle volume in $(\mu m)^3$/cc at 44°40'N, 22-23 June 1972.

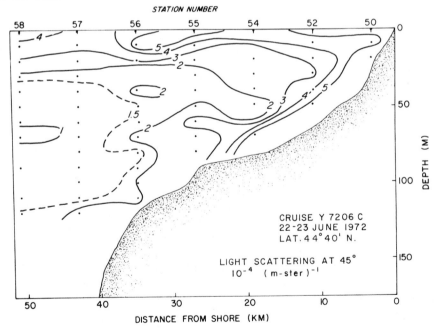

Fig. 4. Light scattering at 45° in $(m\text{-ster})^{-1}$ x 10^{-4} at 44°40'N, 22-23 June 1972.

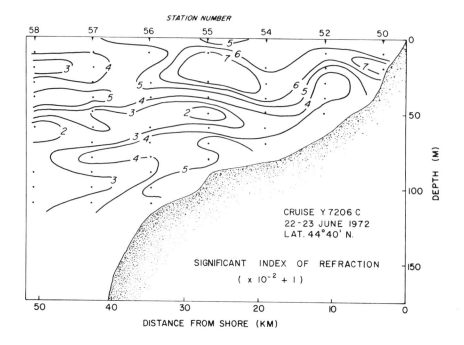

Fig. 5. Difference between particle and water indices-of-refraction x 10^{-2} + 1 at 44°40'N, 22-23 June 1972.

Figures 6, 7, and 8 show the same parameters shown in Figures 3, 4, and 5 but at a location 15 miles to the north. The distribution of the parameters is similar to that 15 miles to the south. Of interest is a tongue of turbid, high-index water extending westward from the bottom at about 70 m depth. This flow is nearly coincident with a surface of constant density. It can be identified as a flow of stirred-up bottom water both by the concentration of suspended matter and its origin; it was not visible in the hydrographic parameters. It should be noted that the flow extends at least 15 miles in a north-south direction and 20 miles in an east-west direction.

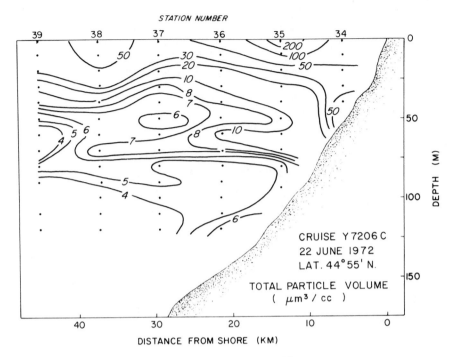

Fig. 6. Total particle volume in $(\mu m)^3$/cc at 44°45' N, 22 June, 1972.

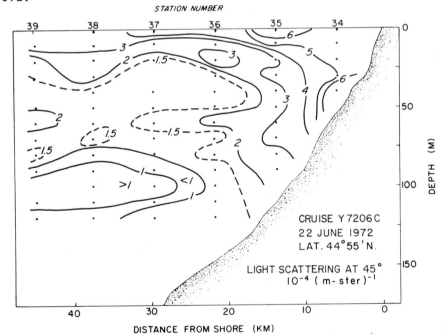

Fig. 7. Light scattering at 45° in $(m$-ster$)^{-1}$x 10^{-4} at 44°45'N, 22 June, 1972.

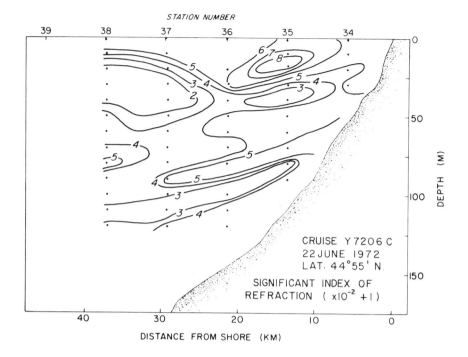

Fig. 8. Difference between particle and water indices-of-refraction x 10^{-2} + 1 at 44°45'N, 22 June 1972.

Figures 9, 10, and 11 show the distribution of the particle parameters at the same locations shown in Figures 3, 4, and 5, but following a period of strong winds. These figures are presented to indicate that the high correlation between particle concentration and index of refraction in the earlier figures is a result of the dynamic structure of the ocean and is not a result of the method used for the calculation of the index of refraction. A comparison of Figure 11 to Figure 5 shows that the stratification of the index of refraction near the surface has been broken due to the influence of increased mixing. The result and distribution of the index of refraction is thus much more uniform.

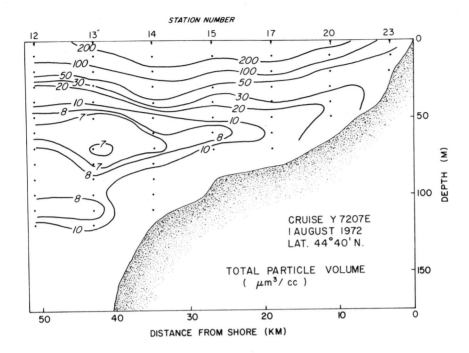

Fig. 9. Total particle volume in (μm)³/cc at 44°40'N, 1 August, 1972.

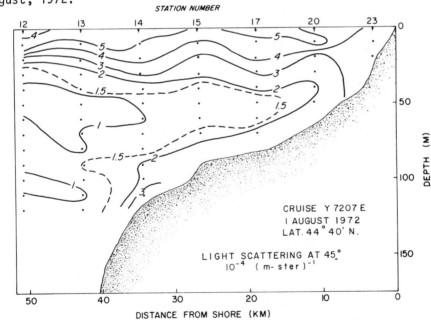

Fig. 10. Light scattering at 45° in (m-ster)⁻¹ x 10⁻⁴ at 44°40'N, 1 August 1972.

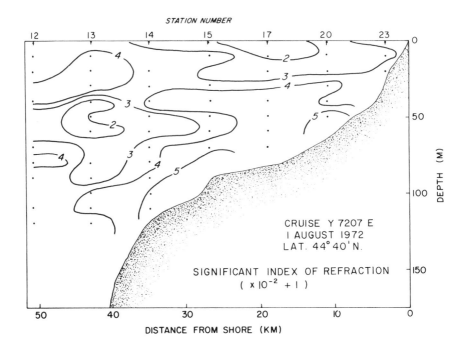

Fig. 11. Difference between particle and water indices-of-refraction x 10^{-2} + 1 at 44°40'N, 1 August 1972.

The index of refraction was also calculated for samples taken near the Galapagos Islands (Figure 12). The interaction of the Galapagos Islands with the Cromwell current, which flows west of the Galapagos eastward along the equator, is described by *Pak and Zaneveld* [in preparation, 1973]. In this region, a wake of high-index particulate matter on the east side of the islands in the Cromwell current would be expected. However, this wake is only weakly indicated by the distribution of total particle count and index of refraction. Away from the islands the particulate matter does not seem to be well-differentiated; the index of refraction fluctuates in a very small range.

DISCUSSION AND CONCLUSIONS

The method used in this paper to estimate the index of refraction of suspended particles is subject to several errors and is to be considered as a first approximation. Continued work in the field will yield more accurate methods that can also be applied routinely to large ocean areas. Preferably, instruments should be built to measure the total scattering coefficient. Combined with improved particle counters these should result in more accurate determinations of the index of refraction of a sample.

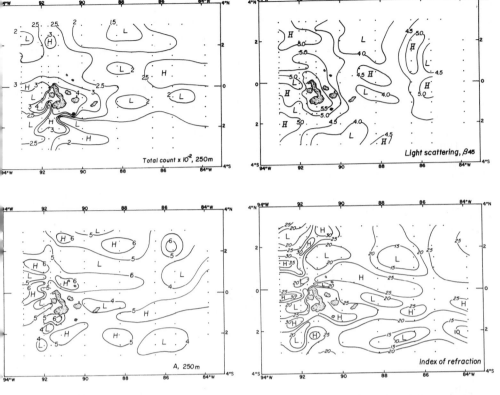

Fig. 12. (a) Total particle count x 10^{-2} (μm)^3/cm^2; (b) light scattering at 45° x 10^{-4} m^{-1}-ster^{-1} particle-size distribution parameter A x $10^{-1}m^{-1}$ and the difference between the (c) particle and (d) water indices-of-refraction x 10^{-3} + 1 at 250 m depth near the Galapagos Islands.

The method currently used shows sufficient differentiation between types of particles to be useful in coastal environments. In open-ocean situations the particulate matter is much more uniform and more accurate methods must be developed if the index of refraction is to be used there as an effective tracer.

ACKNOWLEDGMENTS

The research for this paper received support from the Office of Naval Research.

REFERENCES

Brown, O. B. and H. R. Gordon, Comment on "method for the determination of the index of refraction of particles suspended in the ocean," in preparation, 1973.

Gordon, H. R., H. Bader, and O. B. Brown, An experimental and theoretical study of suspended particulate matter in the Tongue of the Ocean and its influence on underwater visibilities, *Final Report*, Contract No. FО8606-7-C-0028 Air Force Eastern Test Range Measurements Laboratory, 1971.

Jerlov, N. G., Particle distribution in the ocean, *Rep. Swed. Deep-Sea Exped.*, *1947-1948*, vol. 3, pp. 73-97, 1953.

Kullenberg, G., A comparison between observed and computed light scattering functions, *Inst. Fys. Oc U. of Copenhagen, Rep. 13*, 12 + 9 pp., 1970.

Mie, G., Beiträge zur Optik trüber Medien, speziell kolloidalen Metallösungen, *Ann. Physik.*, *25*, 377, 1908.

Morel, A., Diffusion de la lumière par les eaux de mer, Resultats experimentaux et approche théorique, in *Optics of the Sea* (Interface and In-water Transmission and Imaging) *AGARD-LS-61*, AGARD-NATO, pp. 3.1-1 - 3.1-76, 1973.

Pak, H. and J. R. V. Zaneveld, The Cromwell current on the East side of the Galapagos Islands, in preparation, 1973.

Sasaki, T., N. Okami, G. Oshiba, and S. Watanabe, Angular distribution of scattered light in deep sea water, *Records of Oceanographic Works in Japan*, vol. 5, no. 2, pp. 1-10, 1960.

Van de Hulst, H. C., *Light Scattering by Small Particles*, pp. 172-199, Wiley, New York, 1957.

Zaneveld, J. R. V., and H. Pak, Method for determination of the index of refraction of particles suspended in the ocean, *J. Opt. Soc. Am.*, *63*, 321-324, 1973.

Absolute Calibration of a Scatterance Meter

EDWARD S. FRY

Texas A & M University

ABSTRACT

A new method has been developed for determining the absolute calibration of a light-scattering meter. The method is applicable when the light source is a well-collimated beam of small diameter, such as from a laser. No special reflectance or transmittance standards are required and the method is simple to implement. The measured calibration factor includes refractive effects at water/glass/air interfaces as well as the changes in scattering volume with scattering angle.

INTRODUCTION

An important means of studying suspended particulate matter in the ocean is measuring its light-scattering properties. A complete determination of the scattering properties must include their effects on the polarization of the light. In order to completely specify the intensity and polarization characteristics of an arbitrary light beam, four independent parameters are required. A typical set might be the two intensities transmitted by linear polarizers oriented in perpendicular directions, an angle representing the orientation of the polarization ellipse, and the ellipticity of the beam. The usual choice is the four-component Stokes' vector. If a light beam is scattered by suspended matter, then the Stokes' vectors of the incident and scattered beams are related by a four-by-four matrix called the phase matrix. This matrix depends only on the properties of the scattering sample. The Stokes' vector and the phase matrix are discussed fully by *Van de Hulst* [1957].

101

Some components of the phase matrix are particularly sensitive to such properties of the hydrosol as particle orientation and optical activity, the latter being a characteristic of chlorophyll [*Philipson et al.*, 1971] in living plants. These components may, therefore, provide sensitive monitors of biological activity. The phase matrix is also required in calculations of underwater radiance distributions which include polarization effects. The matrix at forward scattering angles is especially important since light propagating through a medium scatters most frequently at small forward angles. There are known cases of polarization sensitivity in the eyes of arthropods and cephalopods [*Waterman*, 1973]; hence, the underwater polarization distributions may have important biological significance. At present, there have been only two measurements of the phase matrix of ocean water; these were by *Beardsley* [1968] and by *Kadyshevich et al.* [1971].

While setting up a program of phase matrix measurements, a new method for absolute calibration of the scatterance meter was developed. The present paper describes the theoretical analysis of this calibration method. It is a method that does not require any special standards and considerably simplifies the calibration procedure — important considerations since the instrument must be calibrated at all sixteen polarization configurations for every angle and wavelength at which it is used. It should be noted that the major factors affecting the calibration are: (*1*) the change in size of the observed scattering volume as the scattering angle is changed, (*2*) refractive effects at the entrance window to the detector system, and (*3*) variations in photomultiplier sensitivity with respect to both polarization and point of incidence on the tube face.

Pritchard and Elliott [1960] devised a scanning technique using a calibration screen to correct for changes in sample volume. This technique forms the basis for the calibration method described by *Tyler* [1963], as well as for the method described here. The method of Pritchard and Elliott also makes determination of input irradiance unnecessary, but requires a reflectance standard. Tyler's method eliminates the need for the reflectance standard, but requires a special procedure for determining absolute reflectance of the scanning screen. The method presented here reduces the calibration procedure to a few simple measurements. It applies to instruments in which the incident light source is a small diameter laser beam.

CALIBRATION THEORY

In the measurement of the volume-scattering function for a water sample, the radiant intensity scattered at angle θ to the incident beam direction by an element of volume dV is

$$dI = \beta(\theta)E dV, \tag{1}$$

where E, the irradiance, is the radiant flux from the laser which is incident on a unit area normal to the laser beam at the position of the volume element dV in the sample cell and $\beta(\theta)$ is the volume-scattering function to be measured. The radiant flux reaching the detector from this volume element is

$$dF = \eta C_1 dI = \eta\beta(\theta)C_1 E dV, \qquad (2)$$

where η is the fraction of the flux transmitted through the cell wall and C_1 is a solid-angle factor. C_1 is zero for dV outside the view of the detector; it also includes the effects of refraction at the cell wall. Figure 1 shows the geometry of C_1 for 3 different points in the scattering volume. This element of flux produces a meter reading

$$dW_w = T_w C_2 dF = \eta T_w \beta(\theta)C_1 C_2 E dV , \qquad (3)$$

where T_w is the transmittance of a neutral density filter placed in front of the detector to keep it in the linear region and C_2 takes into account the sensitivity of the detector and the gain of the associated electronics. In general, C_2 is a function of the position of the element dV since each volume element illuminates a different portion of the detector surface which may have a non-uniform response.

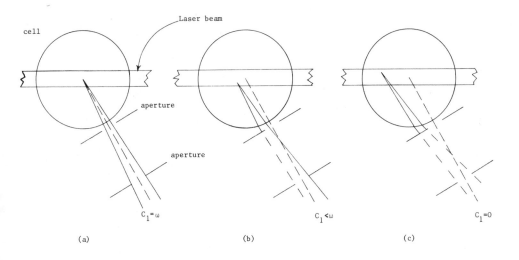

Fig. 1. Geometry of the factor C_1. (a) The sample point is at the center of the scattering volume and $C_1 = w$; (b) The light cone subtended by the final aperture is partially blocked by the first aperture so that $C_1 < w$; (c) The light cone subtended by the final aperture is completely blocked by the first aperture so that $C_1 = 0$.

Integrating (3) over the scattering volume gives the meter reading

$$W_w = nT_w \beta(\theta) \int\int\int_v C_1 C_2 E dV, \tag{4}$$

where θ is the angle between the incident beam direction and the detector axis, and $\beta(\theta)$ is the average value of β over the scattering angles accepted by the detector apertures. This does not introduce appreciable error since β is generally a slowly varying function of θ and the range of scattering angles accepted by the detector is small.

In order to evaluate β from (4), the volume integral must be expressed in terms of measureable quantities. This is the calibration problem. The procedure we have developed is described here.

As in the method of Pritchard and Elliott, a thin, diffusing plastic screen is moved through the sample space and the reflected or transmitted intensity is measured. In this case, the screen is submerged in water and the intensity of the radiation diffused by the screen is measured as the screen moves. The intensity of radiation scattered by the water during calibration is lower by a factor of approximately 10^4 from that scattered by the screen and hence has negligible effect on the calibration. However, the light scattered by the screen is rescattered by the water so that the entire sample cell glows dimly and a uniform background is produced under the calibration signals. This background may be several per cent of the peak calibration signal but is easily subtracted out as discussed later.

A rectangular coordinate system oriented so that the x-axis coincides with the laser beam axis and the y-axis lies in the scattering plane is given in Figure 2. The thin diffusing plastic screen is moved along the x-axis. With an element dA in the cross-sectional area A of the incident laser beam, the radiation passing through dA will be incident on an area $dA' = dA/\cos\gamma'$ of the diffusing screen; γ' is the angle between the normal to the screen and the incident beam. The energy radiated per unit time per unit solid angle from dA' in the direction of the detector when the screen is at a position x is

$$dI_c = \cos\gamma'' \, L_c dA' = \frac{\cos\gamma''}{\cos\gamma'} \, L_c dA, \tag{5}$$

where γ'' is the angle between the normal to the screen and the detector axis and L_c is the appropriately polarized radiance of the screen in the direction of the detector under these conditions. The flux (radiant energy per unit time) reaching the detector from dA' is

$$dF_c = nC_1 dI_c, \tag{6}$$

where C_1 and n are as already defined. C_1 will be zero for dA'

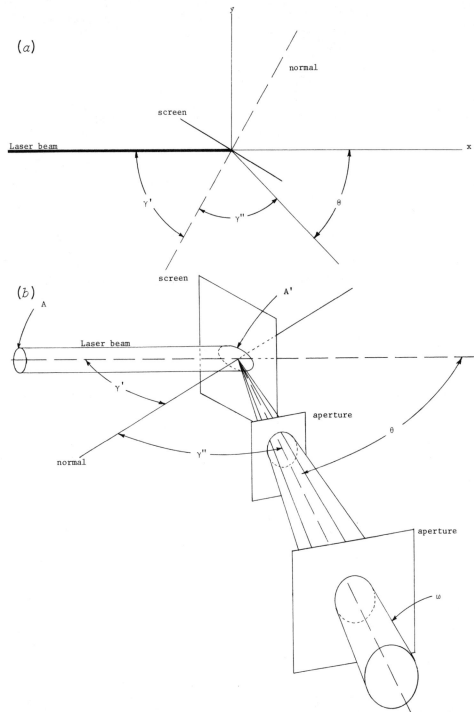

Fig. 2. Schematics showing the diffusing screen at $x = 0$ and illustrating the calibration parameters (a) in the scattering plane and (b) from a perspective.

outside the view of the detector and will depend, in general, on the three coordinates x, y, and z. This element of flux will produce a meter reading

$$dW_c = T_c C_2 dF_c , \qquad (7)$$

where C_2 is as already defined and T_c is the transmittance of the neutral density filter. Combining (5), (6), and (7) and integrating over the cross section of the laser beam yields

$$W_c = nT_c \frac{\cos \gamma''}{\cos \gamma'} \iint_A C_1 C_2 L_c dA . \qquad (8)$$

Typical experimental measurements of W_c versus x are shown in Figure 3. The small slope in the top of the curves is expected and is due to the x dependence of the distance between the detector and the screen. If (8) is integrated with respect to x, we obtain

$$\int_x W_c dx = nT_c \frac{\cos\gamma''}{\cos\gamma'} \iiint_A C_1 C_2 L_c dA dx = nT_c \frac{\cos\gamma''}{\cos\gamma'} \iiint_V C_1 C_2 L_c dV. \qquad (9)$$

This result gives the volume calibration, but L_c must now be

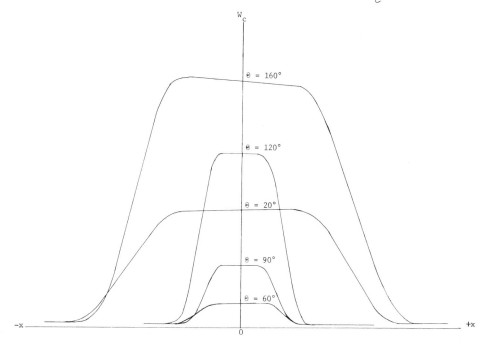

Fig. 3. Measured calibration functions $W_c(x)$. The slight slope of the top of the curves and the apparent shift of the $\Theta = 20°$ curve with respect to the $\Theta = 160°$ curve are real effects of the scattering-volume geometry. W_2 is given by $W_c(0)$.

expressed in terms of readily measured quantities. According to the definition of γ', the polarized radiance of the screen in the direction of the detector at angle θ is

$$L_c = R(\gamma',\theta)E \cos\gamma',\qquad(10)$$

where $R(\gamma',\theta)$ is a reflectance factor for light which has the appropriate polarization to pass through the polarizing elements on the detector axis, and may depend on the polarization of the incident radiation. Its units are reciprocal solid angle. When the screen is at the center of the sample volume, the meter reading will be, by (8),

$$W_2 = nT_c \frac{\cos\gamma''}{\cos\gamma'} \omega \iint_A C_2 L_c \, dA,\qquad(11)$$

where ω is the solid angle subtended by the detector. Since the diameter of the laser beam is small (0.75 mm) compared to detector apertures, the irradiated area of the screen will be very small (only angles γ' for which this is true are used) and will be entirely within the view of the detector. Thus, the solid angle ω can be accurately measured and will be approximately constant for the entire irradiated area of the screen.

   If the exit window is a spherical surface centered at the center of the cell, then ω is the area of the final detector aperture divided by the square of the distance between this aperture and the center of the cell. For other window surfaces, a refraction correction must be made.

   From (10) and (11) we find

$$W_2 = nT_c \cos\gamma'' R(\gamma',\theta)\omega \iint_A C_2 EdA,\qquad(12)$$

where $R(\gamma',\theta)$ is the average value of R over the small range of angles θ accepted by the detector apertures. This introduces negligible error since screen surfaces and angles γ' and θ can be chosen so that R is nearly constant over the desired angular range. It does mean that one should avoid the specular angle during calibration. Note also that $\theta = 180° - \gamma' - \gamma''$.

   When the screen is removed and the detector is placed at $\theta = 0$ to observe the direct beam, the meter reading is

$$W_1 = nT_1 \iint_A C_2' \, EdA,\qquad(13)$$

where n is as defined previously, T_1 is the transmittance of the neutral density filter, and C_2' takes into account the sensitivity of the detector and the gain of the associated electronics. If the detector has uniform sensitivity across its face, then C_2 and C_2' are equal and are constants which may be removed from the integrals. Equations (12) and (13) then give

$$R(\gamma',\Theta) \;=\; \frac{T_1 W_2}{T_c W_1 \,\omega\cos\gamma''} \quad . \tag{14}$$

Using (14) and (10), (9) becomes

$$\int_x W_c dx \;=\; \frac{nT_1}{\omega}\,\frac{W_2}{W_1}\,\iiint_V C_1 C_2 E\,dV. \tag{15}$$

The volume-scattering coefficient is obtained from (4) and (15):

$$\beta \;=\; \frac{T_1 W_2}{T_w \omega W_1 \int_x W_c dx}\,W_w \;=\; KW_w \quad . \tag{16}$$

The absolute calibration factor K is thus given in terms of the following readily measured parameters of the instrument:

ω: The solid angle subtended by the detector at the center of the sample cell.

$W_c(x)$: The meter reading as a function of position x of the calibration screen.

W_2: The meter reading when the calibration screen is at the center of the cell; it can be read from the curve W_c versus x.

W_1: The meter reading at $\theta = 0$ with the calibration screen removed.

T_1, T_w: The transmissions of neutral density filters used when measuring W_1 and W_w respectively.

In deriving (14), it is assumed that the detector response is uniform across its aperture. When this is not the case, averaging techniques must be used to obtain R. The simplest technique presently in practice is to permanently mount a diffuser several millimeters in front of the detector, thus providing a uniform irradiance on the detector regardless of the irradiance distribution incident on the diffuser. The uniformity of reponse of such a system must always be checked experimentally. The principal objection to this technique is the loss of signal from scattering in the diffuser. This problem can be circumvented by noting that the non-uniformity of response introduces difficulties only in the evaluation of R in (14). Hence, the appropriate procedure is to determine R from measurements of W_1 and W_2 with the diffuser in place. The diffuser is then removed, and the calibration is completed by measuring $\int W_c dx$. The signals W_w are also measured without the diffuser and the calibration is then given by (16).

DISCUSSION

Accuracy in the calibration is determined largely by the signal to noise and the drift in the instrumentation. Due to the spatial coherence of the laser beam, there will be speckle noise [Myers and Winns, 1972] superimposed on the calibration functions. When this is a problem, it is easily eliminated by oscillating the

screen in its plane with a small amplitude (on the order of 1 mm).
Other systematics that might affect the calibration were discussed
in the presentation of the procedure.

Perhaps the most outstanding feature of the calibration technique presented here is its operational simplicity. Once ω and the
filter transmittances have been obtained, calibration requires only
measurements of W_1 and $W_c(x)$. The method is independent of the
specific reflectance (transmittance) properties of the screen —
provided only that they are slowly varying functions of angle. For
example, Teflon plumbers' tape and plastic bag materials have been
found to provide useful screens. Futhermore, the screen orientation
does not appear in the calibration. Hence, the same calibration
factor K is measured with different values of γ'.

REFERENCES

Beardsley, G. F., Jr., Mueller scattering matrix of sea water,
    J. Opt. Soc. Am., *58*, 52-7, 1968.
Kadyshevich, YE. A., YU. S. Lyubovtseva, and I. N. Plaklina,
    Measurement of matrices for light scattered by sea water,
    Atmospheric and Oceanic Physics, *7*, 557-61, 1971.
Myers, M. E. and A. M. Winns, Elimination of speckle noise in
    laser light scattering photometry, *Appl. Optics*, *11*, 947-9,
    1972.
Philipson, K. D., S. C. Tsai, and K. Sauer, Circular dichroism
    of chlorophyll and related molecules calculated using a point
    monopole model for the electronic transitions, *Journal of
    Physical Chemistry*, *75*, 1440-1445, 1971.
Pritchard, B. S. and W. G. Elliott, Two instruments for atmospheric
    optics measurements, *J. Opt. Soc. Am.*, *50*, 191-202, 1960.
Tyler, J. E., Design theory for a submersible scattering meter,
    Appl. Optics, *2*, 245-8, 1963.
Van de Hulst, H. C., *Light Scattering by Small Particles*, John
    Wiley & Sons, New York, 1957.
Waterman, T. H., Polarimeters in Animals, in *Planets, Stars, and
    Nebulae*, edited by T. Gehrels, Univ. of Arizona Press,
    Tucson, pp. 472-494, 1973.

III

NEARSHORE STUDIES

Effects of Tropical Storm Agnes on the Suspended Solids of the Northern Chesapeake Bay

JERRY R. SCHUBEL

Johns Hopkins University

ABSTRACT

In the upper reaches of the northern Chesapeake Bay there are two distinctive distributions of suspended sediment and associated patterns of sediment transport. During the spring freshet, the Susquehanna River overpowers the characteristic net non-tidal estuarine circulation in the upper 20-30 km of the estuary and the net flow and sediment transport are seaward at all depths. Generally the bulk — probably 70 to 75 percent — of each year's supply of new fluvial sediment is introduced during the spring freshet when both riverflow and concentration of suspended sediment are normally highest. The marked decrease seaward of the concentration of suspended solids in the upper bay reveals the close link, during the freshet, between the suspended sediment population and the principal "ultimate" source of fluvial sediment—the Susquehanna river.

With subsiding river flow, the net non-tidal estuarine circulation is re-established in the upper reaches of the bay and a turbidity maximum is formed near the head of the estuary. The high concentrations of suspended solids, greater than those either farther upstream in the source river or farther seaward in the estuary, are produced and maintained primarily by the periodic resuspension of bottom sediment by tidal scour and by the sediment trap created in the upper reaches of the estuarine circulation regime.

The passage of tropical storm Agnes in June 1972, resulted in record flooding throughout the drainage basin of the northern Chesapeake Bay. On June 24, the day the Susquehanna crested at its mouth, the instantaneous peak flow exceeded 32,000 m^3/sec. The daily average discharge of 27,750 m^3/sec. for that day

exceeded the previous daily average high by nearly 33 percent.
Throughout the bay, salinities were reduced to levels lower than
any previously observed. On June 26, 1972, salinities were less
than 0.5% from surface to bottom throughout the upper 60 km of the
bay, and the surface salinity was less than 1% in the upper 125 km
of the bay. Salinities remained low throughout most of the summer,
but had nearly recovered to normal levels by September.

On June 24, the concentration of suspended solids at the mouth
of the Susquehanna River exceeded 10,000 mgℓ^{-1} and, in a one-week
period, the sediment discharge exceeded that of the past several
decades. The bulk of this sediment was deposited in the upper
40 km of the bay.

INTRODUCTION

Most of the papers in this symposium volume are concerned
with suspended matter in the open ocean—away from coastal areas.
In the open sea the concentration of suspended solids is always
relatively low, rarely exceeding 1 mg/ℓ; the size distribution
is generally relatively narrow and uniform, and the composition
largely organic. Even in the deep-sea nepheloid layer, concentrations of total suspended matter rarely exceed 0.5 mg/ℓ,
although in the nepheloid layer of some submarine canyons the
concentration may approach 10 mg/ℓ after severe storms [*Drake,*
this volume].

In estuaries and other coastal areas, the concentrations of
suspended solids are generally much higher than in the open sea
and the concentration, size distribution, and composition much
more variable both in time and in space than farther seaward.
In addition to the "normal" variations, marked fluctuations can
result from catastrophic events such as floods or hurricanes.
There are few direct observations of the effects of "rare" events
on the suspended solids populations of coastal areas, or of any
other part of the marine environment. Sampling during such
episodes is generally difficult and the infrequency of such
occurrences makes the likelihood of fortuitous observations very
small. As a result, the sedimentary impact of severe storms is
much more commonly inferred from an after-the-fact examination of
the "record" than from direct observations of sediment dispersal.

In late June of 1972, tropical storm *Agnes* passed through
the drainage basin of the Chesapeake Bay. Little wind was
associated with the storm in this area but torrential rains raised
flows of the major tributaries to record or near-record levels.
The flooding rivers dumped large masses of suspended sediment into
the Chesapeake Bay estuarine system. Within one week the Susquehanna River discharged more suspended solids into the bay than it
probably had during the past half-century. Concentrations of
suspended solids throughout the northern bay soared to levels
higher than any previously recorded [*Schubel and Zabawa,* 1973].

The Chesapeake Bay Institute made extensive observations
to document the impact of tropical storm *Agnes* on the distributions
of suspended sediment, salinity, temperature, dissolved oxygen,
and nutrients in the Chesapeake Bay. Sampling was initiated during
the period of peak flooding and, for nearly a year following, to
document the impact and subsequent recovery of the bay to
"normal" conditions. The primary purpose of this paper is to
summarize some of the suspended-solids data and salinity data
collected in the northern Chesapeake Bay during the storm and in
the two months following it. In order to properly interpret
these data and to assess the impact of *Agnes*, the "normal"
spatial and temporal distributors of these properties must first
be established.

STUDY AREA

The present discussion will be restricted to the northern
Chesapeake Bay area, defined for the purposes of this paper
as that section of the bay north of 38°58'N (Figure 1). The
northern Chesapeake Bay is clearly "the estuary of the Susquehanna."
With a long-term average discharge of approximately 985 m^3/sec,
the Susquehanna River discharges more than 93 percent of the total
fresh water input to this segment of the bay. The Susquehanna
discharges approximately half of the total fresh water input to
the entire Chesapeake Bay estuarine system, and has the largest
discharge of any river entering the Atlantic Ocean through the
United States. The characteristic seasonal variation of its
discharge — high flow in the spring followed by low-to-moderate
river flow throughout the summer and most of the fall — is
typical of mid-latitude rivers (Figure 2).
The flow regime of the Susquehanna and the associated
circulation patterns generated within the upper reaches of the
northern bay in response to the varying role of the river, produce two distinctive distributions of suspended solids and
concomitant patterns of transport of the suspended solids. The
first characterize the spring freshet and other brief periods
of very high flow. The second, characteristic of periods of low-
to-moderate flow, typify most of the remainder of the year. These
characteristic distributions and routes and rates of suspended
sediment dispersal have been described in some detail by *Schubel*
[1968*a*, 1969] and will be only briefly summarized here.

Periods of High River Flow

During the spring freshet and other occasional short periods
of very high river flow, the Susquehanna River dominates the
circulation in the upper reaches of the northern Chesapeake
Bay; the characteristic net non-tidal estuarine circulation is
overpowered and the net flow is seaward at all depths. River
domination of the circulation in the upper 25-35 km of the bay

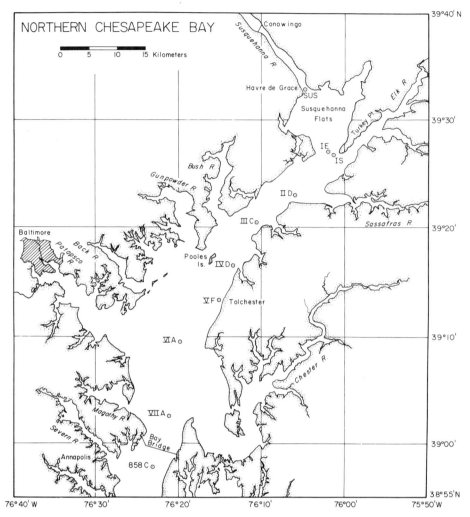

Fig. 1. Map of northern Chesapeake Bay showing station locations and important geographic reference points.

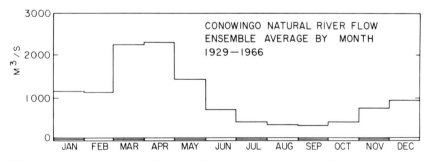

Fig. 2. Susquehanna River flow at Conowingo, Maryland, ensemble average by month, 1929-1966.

during the spring freshet is to be expected, considering the
discharge and the geometry of this segment of the estuary. A
river flow of 3000-4000 m^3/sec would produce a mean seaward
velocity of 12-15 cm/sec through an average cross section upstream
from Pooles Island (39°17'N) and would, in about two days, discharge a volume of water equal to that of the bay upstream from
this island [*Boicourt*, 1969]. Discharge is frequently so great
during the freshet that the tidal reaches of the Susquehanna River—
the stretch of the river above the landward limit of sea salt
intrusion, but still subject to tidal action — extend nearly as
far seaward as Tolchester, almost 45 km from the mouth of the river
at Havre de Grace, Maryland, and 30 km from the head of the
bay at Turkey Point.

During such periods of high flow, the transition from river
to estuary is marked by a sharp front separating the fresh river
water from the saline estuary water (Figure 3). Longitudinal

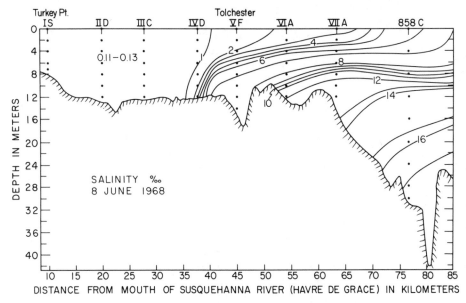

Fig. 3. Longitudinal salinity distribution in northern Chesapeake Bay during a period of high river flow.

salinity gradients larger than 6°/₀₀ in 5 km are common during the
spring freshet. This front moves upstream and downstream in
response to changing river flow but, until June 1972, had not been
observed farther seaward than Tolchester, about 39°13'N. Seaward
of the front where the characteristic net non-tidal estuarine
circulation regime is maintained, stability of the water column
increases with increasing river flow, and vertical mixing decreases. In net estuarine circulation, there is, averaged over
many tidal cycles, a net flow seaward in the upper layer and a
net flow directed up the estuary in the lower layer. This net
non-tidal flow has velocities approximately one-fifth the magnitude

of the maximum tidal currents.

Generally, the bulk — probably 70 to 75 percent by mass — of each year's supply of new fluvial sediment is introduced during the spring freshet when, normally, both river flow and the concentration of total suspended solids are the highest. In most years, the Susquehanna probably discharges 0.5 - 1.0 x 10^6 metric tons of suspended solids into the upper Chesapeake Bay [Schubel, 1968 a, b, 1969; Biggs, 1970]. This sediment is nearly all fine-grained silt and clay; the coarser particles are trapped upstream in the reservoirs along the lower reaches of the river [Schubel, 1968 a, b; 1969].

Except near the bottom, concentrations of total suspended solids greater than 100 mg/ℓ are relatively uncommon in the northern Chesapeake Bay, even in the mouth of the Susquehanna River. In 1967, the maximum concentration in the mouth of the Susquehanna was 140 mg/ℓ [Schubel, 1968b]. In 1969, the maximum concentration of suspended sediment at Conowingo was only 57 mg/ℓ [Schubel, 1972]. The Conowingo Hydroelectric Plant is located at Havre de Grace, Maryland, approximately 15 km upstream from the river's "mouth" and is the last dam before the river enters the bay. Since 1969, samples of suspended solids have been collected on the downstream side of the Conowingo Dam on nearly a daily basis. In 1970, the maximum concentration at Conowingo was 253 mg/ℓ but it exceeded 100 mg/ℓ on only five days during the entire year. In 1971, the peak concentration at Conowingo was 142 mg/ℓ and it exceeded 100 mg/ℓ on only four days during the year.

During the spring freshet there is a steep downstream gradient of the concentration of suspended solids in the upper reaches of the bay [Schubel, 1968a, 1969; Schubel and Biggs, 1969]. In the spring freshet of 1967, the maximum concentrations of total suspended solids in the mouth of the Susquehanna exceeded the maximum concentrations at Tolchester(station V in Figure 4) by as much as a factor of 5. Similar longitudinal gradients have been observed in other years. Simple dilution models based on comparisons of the longitudinal gradients of the concentration of suspended solids and the concomitant longitudinal salinity gradients indicate that more than 70 percent of the sediment discharged during a freshet is deposited upstream from Tolchester (station V) — upstream from the front associated with the encroaching salt water [Schubel, 1968a, 1969]. Biggs [1970] estimated that more than 90 percent of the sediment contributed by the Susquehanna is deposited north of station VI located at 39°09'N.

In the segment of the bay upstream of the salinity front, net flow and sediment transport are downstream (seaward) at all depths. Current measurements made in the upper reaches of the bay during freshets reveal that at all depths the ebb currents predominate over flood currents both in duration and in intensity [Schubel, 1969]. Flood tidal periods are generally of short duration, lasting from 3 to 5 hours, and maximum speeds commonly fall below the critical erosion speeds of the fine-grained

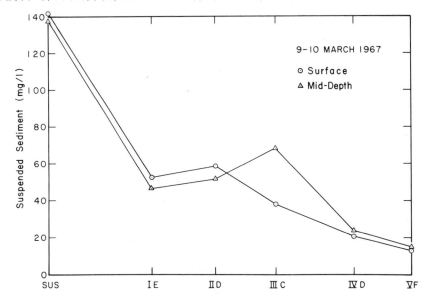

Fig. 4. Surface and mid-depth concentrations of total
suspended solids at the Susquehanna station and at channel stations
within the upper bay during the 1967 freshet.

sediments, 35-50 cm/sec. Ebb periods are much longer, lasting from
7 to 9 hours, and maximum current speeds typically exceed 100 cm/
sec. Bottom sediments, resuspended by the strong ebb currents,
settle out when the current wanes, producing marked fluctuations
in the concentration of suspended sediment. The fluctuations
are related to tidal period since the flood currents are commonly
too weak to erode the bottom. These fluctuations, although
greatest near the bottom, are observed throughout the entire depth
because of the intense vertical mixing through an essentially stable
water column. Salinity, equal to that of the river, is uniform
top to bottom and the temperature gradient is very small. These
conditions are characteristic of the upper 30 km of the bay from
Turkey Point to Tolchester during the spring freshet when the
Susquehanna River dominates the circulation pattern.
     Farther seaward in the estuary — seaward of the front —
where the characteristic net non-tidal estuarine circulation is
maintained, the high fresh water discharge of the freshet results
in increased stability of the water column and decreased vertical
mixing. The vertical distribution of suspended sediment reveals
the action of two sediment sources — river discharge in the
upper layer and the resuspension of bottom sediments by tidal
scour in the lower layer. The fluctuations of the suspended
sediment concentration, produced by tidal "scour and fill," are of
semi-tidal period and are restricted primarily to the lower layer
because the great stability of the water column inhibits vertical
mixing. For a short time at peak flow, the inputs from these two

sources are, in the region of the front, separated by a pronounced
minimum of the concentration of suspended solids in the pycnocline
(Figure 5). This feature, found only in the vicinity of the front

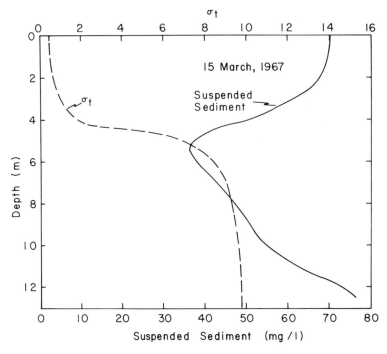

Fig. 5. Total suspended solids (mg/ℓ) and density (σ_t) versus
depth for a station in the frontal region separating the fresh
river water from the salty estuary water during the 1967 spring
freshet.

where there is a very strong vertical density gradient, is a
transitional feature and, of course, cannot persist. A suspended
solids maximum in the pycnocline has never been observed in this
portion of the bay.

Periods of Moderate and Low River Flow

With subsiding river flow, the characteristic net non-tidal
circulation regime is reestablished in the upper reaches of the
Chesapeake Bay. Current measurements averaged over many tidal cycles show, near the surface, an apparent ebb current that is longer
and stronger than the flood current and, near the bottom, an
apparent flood current that is longer and stronger than the ebb
current. Removal of the oscillatory tidal currents from the
observations yields the typical two-layer estuarine flow pattern
[Pritchard, 1952] that characterizes even the upper reaches of the

bay except during the spring freshet and other occasional periods
of very high river flow. With the reestablishment of this
circulation pattern following high runoff, salt is advected into
this portion of the estuary by the lower layer and the salinity
distribution shown in Figure 3 is transformed to resemble that
shown in Figure 6 — the distribution pattern characteristic of

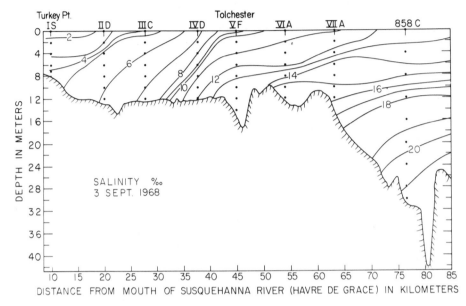

Fig. 6. Longitudinal salinity distribution in the northern
Chesapeake Bay during a period of moderate river flow.

two-layer estuarine circulation regimes. The rate of recovery
following a decline in river flow is not well known but it is
almost certainly less than one week and may be only a few tidal
cycles.

Except for a few days during peak flow of the spring freshet
and other occasional brief periods of very high river flow, the
concentrations of suspended solids are greater within the upper
25-30 km of the estuary than farther upstream in the source river
in spite of both the dilution of Susquehanna inflow and the
settling-out of the sedimentary particles that occur within the
upper bay. At all times of the year, the concentrations of
suspended solids within the upper 20-30 km of the bay are greater
than those farther seaward in the estuary. Such zones of high
suspended sediment concentration, reported in the upper reaches
of many estuaries throughout the world, are called "turbidity
maxima."

A longitudinal distribution of suspended sediment typical of periods of low-to-moderate river flow is shown in Figure 7. The

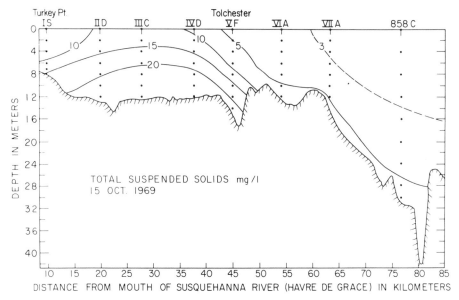

Fig. 7. Longitudinal distribution of total suspended solids
(mg/ℓ) in the northern Chesapeake Bay typical of periods of low-
to-moderate river flow.

steep longitudinal gradient of the suspended solids concentration
between stations IV and V (Tolchester) marks the seaward boundary
of the turbidity maximum. The high concentrations of suspended
solids in the upper reaches of the estuary which persist throughout
the year can not be explained by a gradual purging out of the sediment-laden freshet water since the renewal time is only a few
weeks or less. Nor can the "anomalous" concentrations be explained
by either flocculation [*Lüneburg*, 1939; *Ippen*, 1966] or deflocculation [*Nelson*, 1959] of fluvial sediment [*Schubel*, 1971a].
     Schubel [1968a,c; 1971a] has shown that the turbidity maximum
in the upper reaches of the Chesapeake Bay is produced by a combination of physical processes — periodic local resuspension of
bottom sediments by tidal scour and the "sediment trap" produced
by the net non-tidal estuarine circulation which entraps much of
the sediment, both resuspended and newly introduced, within this
segment of the bay. Maximum tidal currents in the upper bay average
more than 50 cm/sec, while the critical erosion speed of the
sediments away from the littoral zone fall below this value. In
the upper reaches of the bay, concentration of total suspended
solids at 1.5 m above the bottom typically fluctuates,by a factor
as great as seven,between times of slack water and times of
maximum ebb and flood current velocities. At 0.5 m above the bottom,

the concentration typically exhibits a fifteen to twentyfold
fluctuation with a semi-tidal period; variations of 15 mg/ℓ to
300 mg/ℓ are representative. The oscillatory tidal currents
produce marked changes not only in the concentration of suspended
solids but also in their size distribution [*Schubel*, 1971*b*]. At
0.5 m above the bottom the volume-weighted mean Stokes' diameter
commonly exhibits a range of from < 3μ near slack water to > 20μ
near times of maximum ebb and flood currents.

Much of the sediment, both that resuspended and that
newly introduced, is trapped within the upper 30-40 km of the
northern bay by the net non-tidal estuarine circulation. An
effective sediment trap is formed near the head of the estuary
where the net non-tidal upstream flow of the lower layer dissipates
until, finally, the net flow is downstream at all depths.
Particles that settle out of the seaward-flowing upper layer into
the lower layer are carried back upstream by its net non-tidal
upstream flow; sediment then accumulates and a "turbidity maximum"
forms near the head of the estuary [*Postma*, 1967; *Schubel*, 1968*a, c*].
The net non-tidal circulation not only effectively entraps much of
the sediment introduced directly into this segment of the bay but
also supplements it with sediment previously carried through this
segment during periods of high river flow and with sediment introduced from other sources into more seaward segments of the estuary.

Many of the particles suspended in the lower layer are
transported back into the upper layer by vertical mixing; the process
is repeated many times. Mixing, as defined here, includes both
vertical advection and diffusion. Continuity requires that the
water flowing up the estuary in the lower layer be returned
seaward in the upper layer; hence, there must be a vertical
advection of water from the deeper layer into the surface layer.
The speed of this net vertical flow is zero at both surface and
bottom and reaches a maximum speed of 10^{-3} cm/sec near mid-depth
[*Pritchard*, 1956]. In addition, a vertical diffusion velocity
of 10^{-3} cm/sec exists due to turbulence. Within the turbid zone
the mixing during much of the year is of sufficient intensity to
overcome the vertical stratification and to produce a nearly
homogeneous water column twice during each tidal cycle. Farther
seaward, however, vertical mixing is inhibited and the water column
remains stratified over much longer time scales.

1972, THE YEAR OF *AGNES*

The year 1972 began not very unlike most years, although it
was somewhat wetter. During the spring freshet in March, river
flow was fairly high, exceeding 8900 m^3/sec, and the concentration
of total suspended solids in the Susquehanna River at Conowingo
reached 190 mg/ℓ. Between 1 January 1972 and 21 June 1972 the
concentration of suspended solids at Conowingo exceeded 100 mg/ℓ
on only four days — not unlike most years. During May and the
first 20 days of June of 1972, the concentration of total suspended
solids at Conowingo was generally between 10-25 mg/ℓ, somewhat high-

er than the average for this time of year but not really "abnormal."
When tropical storm *Agnes* entered the area, torrential rains fell
throughout most of the drainage basin of the Chesapeake Bay estuarine system. Total rainfall accompanying the passage of the storm
ranged from 20 to more than 45 cm over most of the Susquehanna River drainage basin.

The heavy rains produced record flooding of the Susquehanna.
On 24 June 1972, the day the river crested, the average daily flow
exceeded 27,750 m^3/sec — the highest average daily flow ever recorded, exceeding the previous daily average high by approximately
33 percent. The instantaneous peak flow on 24 June 1972 of more
than 32,000 m^3/sec was the highest instantaneous flow reported over
the 185 years of record (Figure 8). The monthly average discharge
of the Susquehanna of about 5100 m^3/sec or June 1972 was the highest average discharge for any month over the past 185 years, and
was more than nine times the average June discharge over this same
interval. A comparison of the monthly average discharge of the Susquehanna during 1972 and the ensemble monthly average over the period 1929-1966 clearly shows the departure of the June 1972 flow from
the long-term average June flow (Figure 9).

Even before the storm *Agnes*, 1972 had been a "wet" year (Figure 9), and salinities throughout much of the bay were lower than
their more normal values. With the large influx of fresh water
following *Agnes*, salinities fell sharply. The lag between time of
maximum discharge and the time of minimum salinity varied, of course,
with location and depth. In the surface layers of the upper 180 km
of the estuary the salinities reached minimum values within 2-5 days
of the cresting of the Susquehanna. In the near-bottom waters in
the same region, minimum salinities were not reached in some areas
until 14-15 July 1972, 20 days after cresting. The tidal reaches
of the Susquehanna were pushed seaward more than 80 km from the
mouth of the river at Havre de Grace, that is, nearly to the Chesapeake Bay bridge at Annapolis, Maryland. The front, separating
the fresh river water from the salty estuarine water, was more than
35 km farther seaward than ever previously reported (Figure 10).

Reestablishment of the "normal" salinity distribution is effected by the flow of more saline waters up the estuary in the lower
layer and subsequent slow vertical mixing of the lower and upper
layers. The combination of large fresh water inputs accompanying
Agnes and the compensating upstream flow of salty water in the lower
layer produced vertical salinity gradients larger than any previously recorded throughout much of the Chesapeake Bay estuarine system.
Abnormally large gradients persisted throughout the summer. Even
in early autumn the vertical salinity gradients were more typical
of spring conditions than those characteristic of the fall season.

The flooding rivers dumped large masses of sediment into the
Chesapeake Bay estuarine system. Even before *Agnes*, river flow and
the concentration of suspended solids were somewhat higher than normal for that time of year (Figure 11). On 22 June 1972 river flow
increased rapidly as a result of heavy rainfall accompanying tropical storm *Agnes* and the concentration of suspended solids at Conowingo reached 400 mg/ℓ (Figure 11). On 23 June 1972 river flow

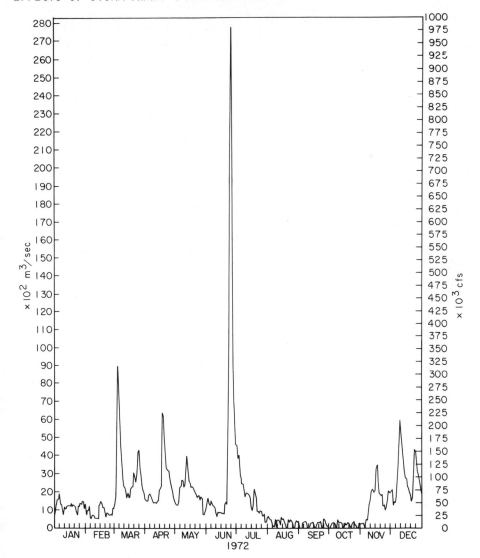

Fig. 8. Discharge of the Susquehanna River at Conowingo, Maryland, during 1972.

exceeded 24,400 m^3/sec and the concentration of suspended solids jumped to more than 10,000 mg/ℓ — a concentration more than 40 times greater than any previously reported for the lower Susquehanna. This concentration was determined on the downstream side of the dam at Conowingo and represents the concentration of suspended sediment in the flood waters being discharged into the upper Chesapeake Bay. Unfortunately, no sediment sample was collected on 24 June 1972, the day the Susquehanna crested, because the area of the dam was evacuated for reasons of safety. The average daily flow on June 24 was

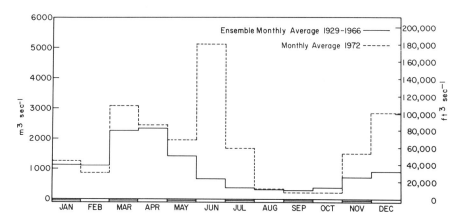

Fig. 9. Ensemble monthly average river flow at Conowingo, Maryland during the period 1929-1966, and the monthly average discharge
for 1972.

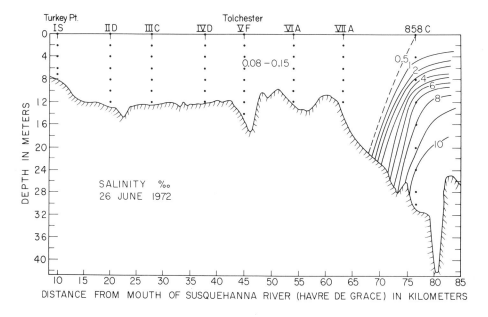

Fig. 10. Longitudinal salinity distribution in the northern
Chesapeake Bay on 26 June 1972 following tropical storm *Agnes*.

more than 27,750 m³/sec. By June 25 the river flow had decreased
to about 23,100 m³/sec and the concentration of suspended solids
had fallen to 1,450 mg/ℓ. By 30 June 1972, river flow had subsided
to about 4,600m³/ sec and the concentration of suspended solids had
subsided to approximately 70 mg/ℓ. During the one-week period
22-28 June 1972, the Susquehanna River probably discharged >50 x 10⁶
metric tons of suspended sediment into the upper Chesapeake Bay

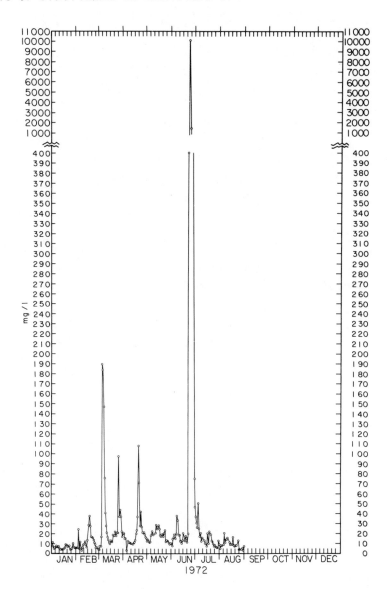

Fig. 11.   Concentration of total suspended solids (mg/ℓ) in the Susquehanna River at Conowingo, Maryland, during 1972.

— more suspended sediment than had been discharged during the past three decades, and probably during the past half-century. The bulk of this sediment, probably more than 85 percent of it by mass, was silt and clay, the remainder was fine sand.
    A large fraction of the sediment discharged into the upper Chesapeake Bay by the Susquehanna was material eroded from the bottom of the Conowingo Reservoir and perhaps from one or more of

the other reservoirs between Conowingo, Maryland, and Harrisburg, Pennsylvania. Cannibalism of the reservoirs may have accounted for as much as 75 or 80 percent of the Susquehanna's total sediment discharge into the upper bay during *Agnes*. Comparison of our data for Conowingo and data from the U. S. Geological Survey [*A. B. Cummings*, personal communication, 1972] for Harrisburg indicates that, between 22-28 June 1972, the suspended solids discharge of the Susquehanna past Conowingo exceeded that at Harrisburg by nearly a factor of seven. The river flow increased by less than 10 percent over the same approximately 100-km stretch of the river.

The Conowingo Reservoir was resurveyed in late summer and early fall 1972, but the sounding records have not been analyzed and it is unlikely that the results will be available before another six to twelve months [*A. Hogan*, Philadelphia Electric Company, personal communication, 1973]. These data will be valuable in estimating the amount of sediment removed from the reservoir by the *Agnes* floodwaters. A significant amount of the sediment discharged by the Susquehanna was deposited below Havre de Grace where the Susquehanna opens into the broad, shoal region known as the Susquehanna Flats (Figure 1). Approximately 10 acres of new islands, and several hundred acres of new inter-tidal areas were formed on the flats during *Agnes*. In addition, more than 38,000 m^3 of new fill had to be dredged from one section of the main shipping channel to restore it to its original project depth.

Following the flooding, the Chesapeake Bay Institute took a large number of core samples in the northern Chesapeake Bay. The cores are being analyzed to determine the thickness of the layer of sediment resulting from *Agnes* at various points in the bay. From these data estimates will be made of the depositional patterns of the sediment resulting from *Agnes*. The data will also provide an independent estimate of the total mass of sediment discharged by the Susquehanna during the flooding.

The large influxes of suspended sediment by the Susquehanna and other rivers produced anomalously high concentrations of suspended solids throughout much of the Chesapeake Bay estuarine system. In the northern bay concentrations soared to levels higher than any previously reported. On 25 June 1972, the concentration of suspended solids off Pooles Island exceeded 800 mg/ℓ at the surface and 1100 mg/ℓ at mid-depth (6 m). Near the time of peak flooding there was a marked longitudinal gradient of the concentration of suspended solids in the northern bay. On 26 June 1972, two days after the Susquehanna crested, the concentration of suspended solids at the surface dropped from more than 700 mg/ℓ at Turkey Point at the head of the bay to about 400 mg/ℓ at Tolchester (30 km farther seaward), and to approximately 175 mg/ℓ at the Annapolis Bay Bridge, 65 km from the head of the bay (Figure 12). The concentration of suspended solids at mid-depth showed a similar distribution pattern although the concentrations were generally greater than near the surface (Figure 12). Seaward

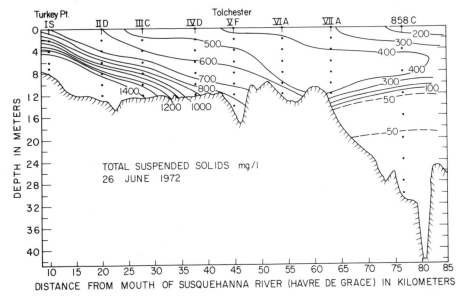

Fig. 12. Longitudinal distribution of total suspended solids
(mg/ℓ) in the northern Chesapeake Bay on 26 June 1972 following
tropical storm *Agnes*.

of station VIIA, however, there was an abrupt decrease of the
concentration of total suspended solids below about 10 m. This
distribution results from the over-riding of the relatively "clean"
estuary water by the sediment-laden Susquehanna water. The
concentrations of total suspended solids remained anomalously high
throughout most of the northern bay for nearly a month (Figure 13).
By the end of July, the distribution was near "normal" for that time
of year.

As the normal two-layered circulation pattern was reestablished
throughout the upper bay there was a net upstream movement of
sediment suspended in the lower layer. Sediment previously
carried downstream and deposited by the flooding waters from
Agnes was resuspended by tidal currents and gradually transported
back up the estuary. The routes of sediment dispersal are clear
but the rates of movement are obscure. The data do not permit
reliable estimates of the rates of sediment transport, particularly
during the recovery period.

Sediments are, of course, the archenemy and ultimate conqueror
of every estuary. Estuaries are ephemeral features on a geological
time scale and are rapidly filled with sediments. As an estuarine
basin is filled, the intruding sea is displaced seaward and the
estuarine basin is gradually transformed back into a river valley
system. Tropical storm *Agnes* presented investigators with an
unprecedented opportunity to document the relative importance of a
catastrophic event in the lifetime of an important estuary. It
appears now that during the one-week period following *Agnes*, the
Chesapeake Bay estuarine system "aged," geologically, by approximately

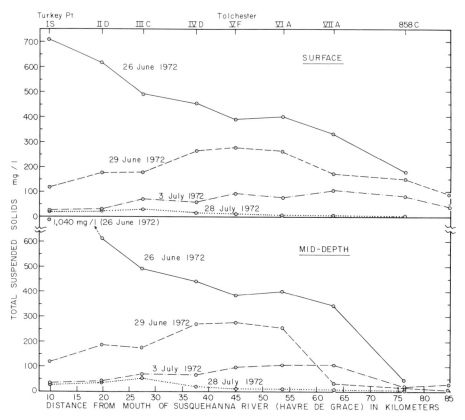

Fig. 13. Longitudinal variations of the surface and mid-depth concentrations of total suspended solids (mg/ℓ) in the northern Chesapeake Bay during several weeks following the passage of tropical storm *Agnes*.

half a century.

SUMMARY

During the spring freshet and other occasional short periods of very high river flow, the upper reaches of the Chesapeake Bay behave like the tidal reaches of a river. The Susquehanna over-powers the characteristic net non-tidal estuarine circulation and the net flow and sediment transport are seaward at all depths. The transition from river to estuary, sometimes as far as 40-45 km seaward of the mouth of the Susquehanna at Havre de Grace, is characterized by a front separating the fresh river water from the saline estuary water. Generally, most of each year's supply of new fluvial sediment is discharged during the freshet. The bulk of this is deposited in the upper 30 km of the bay between cross sections I and V. The spring freshet, then, is a period of fluvial domination of the upper bay and of its suspended sediment population and is characterized by a close link between the suspended sediment population and the principal "ultimate " source of sediment — the Susquehanna River.

With subsiding river flow, the characteristic net non-tidal estuarine circulation is reestablished in the upper reaches of the bay. The concentrations of suspended sediment are greater than those either farther upstream in the source river or farther seaward in the estuary. This zone of high suspended sediment concentration, the "turbidity maximum," is produced and maintained by the periodic resuspension of bottom sediment by tidal scour and by the sediment trap produced by the net non-tidal circulation. No evidence has been found of either flocculation or deflocculation.

The passage of tropical storm *Agnes* in June 1972 resulted in record flooding throughout the drainage basin of the northern Chesapeake Bay. On June 24, the day the Susquehanna crested at its mouth, the instantaneous peak flow exceeded 32,000 m^3/sec. The daily average discharge of 27,750 m^3/sec for that day exceeded the previous daily average high by nearly 33 percent. Throughout the bay, salinities were reduced to levels lower than any previously observed. On 26 June 1972, salinities were less than 0.5°/$_{oo}$ from surface to bottom throughout the upper 60 km of the bay and the surface salinity was less than 1°/$_{oo}$ in the upper 125 km of the bay. Salinities were depressed throughout most of the summer but had nearly recovered to normal levels by September.

On June 24, the concentration of suspended sediment in the mouth of the Susquehanna exceeded 10,000 mg/ℓ and in a one-week period the sediment discharge exceeded that of the past several decades. The bulk of this was deposited in the upper 40 km of the bay.

ACKNOWLEDGMENTS

I am indebted to C. F. Zabawa, W. B. Cronin, T. W. Kana, M. Glendening, and C. H. Morrow for their help both in the field and in the laboratory. This research was supported, in part, by a project jointly funded with Dingell-Johnson funds by the Maryland Department of Natural Resources and the U. S. Bureau of Sport Fisheries and Wildlife; in part, by the Oceanography Section, National Science Foundation, NSF Grant GA-36091; and, in part, by the U. S. Army Corps of Engineers, Baltimore and Philadelphia Districts. This paper is contribution 195 of the Chesapeake Bay Institute.

REFERENCES

Biggs, R. B., Sources and distribution of suspended sediment in
    northern Chesapeake Bay, *Marine Geology*, *9*, 187-201, 1970.
Boicourt, W., A numerical model of the salinity distribution in
    upper Chesapeake Bay, *Chesapeake Bay Institute Technical Rept. 54*,

Johns Hopkins University, 69 p, 1969.

Ippen, A. T., Sedimentation in estuaries, in *Estuary and Coastline Hydrodynamics*, edited by A. T. Ippen, McGraw-Hill, New York, pp. 648-672, 1966.

Lüneburg, H., Hydrochemische Untersuchungen in der Elbmündung mittels Elektrokolorimeter, *Arch. Dtsch. Seewarte, 59*, 1-27, 1939.

Nelson, B. W., Transportation of colloidal sediment in the fresh water marine transition zone (abstract), *1st Internat. Oceanogr. Congr. Preprints*, Amer. Assoc. Adv. Sci., Washington, D. C., 640-641, 1959.

Postma, H., Sediment transport and sedimentation in the estuarine environment, *Estuaries*, edited by G. H. Lauff, pp. 158-179, *Pub. No. 83*, Amer. Assn. Adv. Sci., Washington, D. C., 1967.

Pritchard, D. W., Estuarine hydrography, *Advances in Geophysics*, vol. 1, edited by H. E. Landsberg, pp. 243-280, Academic Press, New York, 1952.

Pritchard, D. W., The dynamic structure of a coastal plain estuary, *J. Mar. Res., 15*, 33-42, 1956.

Schubel, J. R., Suspended sediment of the Northern Chesapeake Bay, *Chesapeake Bay Institute Tech. Rept. 35, Ref. 68-2*, Johns Hopkins University, 264 p, 1968a.

Schubel, J. R., Suspended sediment discharge of the Susquehanna River at Havre de Grace, Md., during the period April 1 1966 through March 31 1967, *Chesapeake Sci., 9*, 131-135, 1968b.

Schubel, J. R., The turbidity maximum of the Chesapeake Bay, *Science, 161*, 1013-1015, 1968c.

Schubel, J. R., Distribution and transportation of suspended sediment in upper Chesapeake Bay, *Chesapeake Bay Institute Tech. Rept. 60, Ref. 69-13*, Johns Hopkins University, 29 p, 1969.

Schubel, J. R., Some notes on turbidity maxima, in *The Estuarine Environment: Estuaries and Estuarine Sedimentation, Short Course Lecture Notes*, edited by J. R. Schubel, pp. VIII3-VIII28, Amer. Geol. Inst., Washington, D. C., 1971a.

Schubel, J. R., Tidal variation of the size distribution of suspended sediment at a station in the Chesapeake Bay turbidity maximum: *Netherlands Jour. Sea. Res., 5(2)*, 252-266, 1971b.

Schubel, J. R., Suspended sediment discharge of the Susquehanna River at Conowingo, Maryland, during 1969, *Chesapeake Science, 13*, 53-58, 1972.

Schubel, J. R., and R. B. Biggs, Distribution of seston in upper Chesapeake Bay: *Chesapeake Science, 10*, 18-23, 1969.

Schubel, J. R., and C. F. Zabawa, A preliminary assessment of the physical, chemical, and geological effects of tropical storm *Agnes* on upper Chesapeake Bay (abstract), *Geol. Soc. America, NE Sectional meeting, Allentown, Pa.*, 1973.

Distribution and Transport of Suspended Particulate Matter in Submarine Canyons Off Southern California

DAVID E. DRAKE

National Oceanographic and Meteorological Laboratories

ABSTRACT

Studies of the distribution of suspended particulate matter in submarine canyons off southern California are reported. Simultaneously collected light beam transmission and salinity/ temperature data demonstrate an association of particle maxima with the relatively steep segments of the vertical, water-density gradient over shelf, slope, and canyon environments. Within the submarine canyons that cut the mainland shelf, mid-water particle maxima contain high percentages of terrigenous detritus supplied by seaward flow from canyon nepheloid layers.

The concentrations of particulate matter in the canyon nepheloid layers rarely, if ever, exceed 10 mg/ℓ even in areas close to major rivers; typical peak values in the canyon heads are 3 to 6 mg/ℓ. It is concluded that turbid-layer, density-excess underflows are not possible at these low concentrations. Further, the suggestion of an earlier report that the slow, net downcanyon transport in Pacific coast canyons could be driven by the suspended particles is not supported by our most recent results from Santa Cruz Canyon. Bottom currents in this canyon are vigorous and show a net downcanyon flow that is far too strong to be explained by turbid-layer flow.

While submarine canyons off southern California are an important pathway for fine sediment leaving the shelf, it is likely that no more than 10-15% of the annual terrigenous supply moves to deeper water by way of the canyons.

INTRODUCTION

Much of the silt and clay-size terrigenous detritus in the sea is introduced by rivers and transported principally as suspended

load.   Understanding of the modes and pathways of particle
transport to the deep sea is of considerable importance to a
wide variety of marine problems.   The complexity of the shelf
environment and the difficulties involved in studies of suspended
matter have, however, impeded such an understanding.

For several years we have been investigating the distribution
and transport of suspended particulate matter over the southern
California borderland in an attempt to shed some light on the
processes involved in seaward transport of this material [*Drake*,
1972; *Drake et al.*, 1972; *Drake and Gorsline*, 1973].

The nearshore zone off southern California can be divided
geomorphologically into five littoral cells, each beginning at a
rocky headland and ending at a submarine canyon [*Inman and
Chamberlain*, 1960; *Ingle*, 1966].   Although *Trask* [1955] and
Gorsline [1971] have shown that some leakage of sand occurs around
the headlands, the bulk of the coarse-grained detritus delivered
by streams within each cell is trapped in submarine canyons and
eventually moved to the canyon fans.   The textures of sediments
filling the shallow canyon heads reflect this process but contain
large amounts of medium to coarse silt and very fine sand-sized
mica [*Shepard and Dill*, 1966].   In fact, sediments in axes of
canyons off southern California can be characterized by their
high contents of mica and marine plant debris.   Studies by
Vernon [1966] and *Cook and Gorsline* [1972] demonstrate that
much of this material is moved seaward of the surf zone, as a
surficial layer, by longshore and nearshore currents.   Furthermore,
frequent measurements of the sediment fill in Scripps Canyon show
that periods of rapid deposition and subsequent slump failure
correlate well with storms and large waves [*Shepard and Dill*, 1966].

Recently, several authors have argued that fine silt and clay
detritus are also trapped in the mainland shelf canyons and that
the canyons are the principal pathway of suspended particles to
the deeper basins of the borderland.   In particular, *Moore* [1969]
has suggested that terrigenous sediment moves parallel to the coast
as a near-bottom turbid layer and is intercepted by the submarine
canyons at the downcoast end of each coastal cell.   Further, Moore
believed that particle concentrations within the canyons would be
sufficient to generate "low-density" turbidity currents.

For the past two years, we have been concerned with the
sediment transport in submarine canyons and have completed several
joint surveys of canyon-controlled currents and suspended
sediment with F. P. Shepard of Scripps Institution of Oceanography.
The present report is based primarily upon surveys of Hueneme
and Redondo Canyons, which incise the mainland shelf, and
Santa Cruz Canyon which cuts the southern slope of the northern
Channel Islands ridge (Figure 1).

METHODS

Adequate description of the distribution of suspended particles

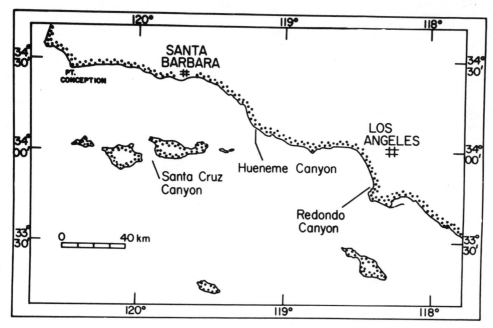

Fig. 1.   Location map for Hueneme, Redondo, and Santa Cruz Canyons.

in the sea can be achieved only with a sampling device that yields real-time data to aid in efficient placement of water sampling equipment. A continuously recording beam transmissometer designed and constructed by the Visibility Laboratory of Scripps Institution of Oceanography [*Petzold and Austin*, 1968] has been used in this study.

This instrument, described by *Tyler et al.* [this volume], measures the volume-attenuation coefficient of water over a 1-m path. This optical variable is controlled by the absorption and scattering of light by the water, suspended particles and dissolved substances [see *Jerlov*, 1968]. Since the transmissometer values are influenced by factors other than particle characteristics, care must be taken to demonstrate that the data principally reflect the distribution of particulate matter. Figure 2 shows the relationship obtained in this study between light transmission values and particle concentrations determined by filtration of samples through Millipore HA discs (nominal pore size: 0.45µ). An analytical precision of $\pm$ 20% was achieved by using two stacked filters for each sample with the lower filter providing a blank correction. Filters were washed three times with approximately 20 mℓ of distilled water, dried overnight at 60°C, and weighed to the nearest 0.01 mg on a Mettler microbalance. Water samples used to construct Figure 2 were obtained from the surface to approximately 600 m in Redondo, Newport and Hueneme Canyons off southern California. All sample stations were within 40 km of the coast and were occupied in the fall and winter months of 1972 and

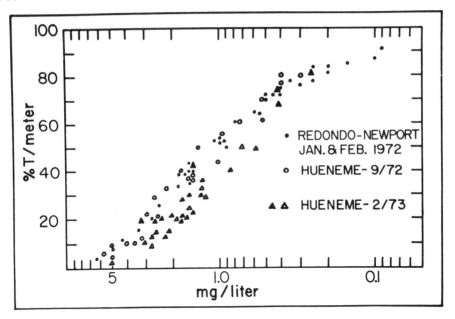

Fig. 2. Scatter diagram of % light transmission versus total suspended particle concentrations (mg/ℓ). The Hueneme Canyon values are displaced 10 to 15% owing to particle size and effects of dissolved organic material on light transmission. Open triangles show samples taken within 15 km of the coast, principally from surface waters, during February, 1973; closed triangles show samples from depths > 50 m or > 25 km from shore.

1973. The good correlation exhibited by the above variables (Figure 2) indicates that (1) concentrations of dissolved, light-attenuating substances were low and relatively uniform during our surveys and (2) with the exception of the 1973 Hueneme data, there were no extreme variations in particle characteristics or dissolved organic concentrations. The 1973 Hueneme survey immediately followed a 4-day period of moderate rainfall and river runoff in southern California. Hueneme Canyon forms the terminus of the first littoral cell which begins at Point Conception 120 km to the northwest and includes the Santa Clara River (Figure 1).

The samples retrieved from depths > 50 m or > 25 km from shore (solid triangles in Figure 2) may be considered to be relatively unaffected by the river discharge. Accordingly, the higher light attenuation for the nearshore samples (open triangles in Figure 2) was probably the result of increases in mean particle size and in concentrations of river-borne dissolved organic material [Jerlov, 1968].

All of the canyon surveys reported here were scheduled during the fall and winter months of 1972 and 1973 for two reasons: to avoid plankton blooms and in order to sample during heavy weather conditions. Plankton blooms typically are pronounced in May and

June off southern California [*Oguri and Kantor*, 1971] and
standing crops are relatively low through the remainder of the
year. A selection of samples from each canyon area was filtered
and the particulate matter was combusted at 450°C for 6 hours
(Table 1). Combustible fractions for surface water samples
ranged from 4% to 26% with the highest values at the seaward limits
of each area (∿ 20 km offshore). These results agree with earlier
work in Santa Barbara Channel [*Drake*, 1972] and with suspensate

TABLE 1. Composition of Suspended Sediments of Hueneme and Redondo
Canyons

	Depth(m)	Total Suspensate (mg/ℓ)	% Combustible Organic Material
Hueneme*			
17912	0/90	2.3/3.2	8/11
17913	0/120	2.1/3.4	12/7
17914	0/161	2.4/3.0	6/10
17915	0/217	1.8/2.4	7/9
17916	0/273	0.8/2.1	13/4
17917	0/329	1.5/1.3	12/6
17918	0/390	1.2/0.6	17/10
17919	0/425	0.9/0.7	15/8
17920	0/510	0.3/0.5	21/11
18456	0/104	2.7/4.2	12/6
18457	0/162	1.8/3.5	10/7
18458	0/195	2.1/2.1	8/8
18461	0/300	1.4/1.2	14/6
18462	0/362	1.5/0.9	10/9
18464	0/472	1.3/0.6	13/12
Redondo*			
16536	0/612	0.31/0.24	26/14
16554	0/590	0.47/0.52	24/18
16551	0/458	0.63/0.91	8/12
16549	0/310	0.53/2.2	16/10
16546	0/68	0.72/1.3	12/12

*Refer to Figs. 3 and 4 for station locations.

composition trends reported for the eastern coast of the United
States [*Manheim et al.*, 1970].
      Sampling during storm conditions was not successful. However,
our February 1973 survey of the Hueneme Canyon area immediately
followed a four-day period of moderate coastal rainfall, approximately

10 cm of precipitation in the period prior to February 14, 1973.
Southern California rivers have essentially intermittent flow with
significant discharge during late fall, winter and spring only.
Much of the Santa Clara River basin is farmland and, therefore,
yields relatively large amounts of suspended matter following coastal storms. Although sediment discharge data collected by the U.S.
Geological Survey during February 1973 are not yet available, statistics for previous years suggest that comparable amounts of rainfall and river flow would yield between 50,000 and 200,000 metric
tons of suspended particulate matter. In contrast, river flow and
sediment discharge had been negligible for 3 months before our
September 1972 survey of Hueneme canyon and the adjacent Ventura
shelf.

STUDY AREAS

To date, we have collected suspended sediment and bottom current data in Hueneme, Redondo, Newport, La Jolla and Santa Cruz
Canyons, all located within the southern California borderland
[see *Emery*, 1960]. Detailed discussions of the results of the studies of Redondo, Newport and La Jolla Canyons are reported elsewhere
[*Drake and Gorsline*, 1973; *Shepard and Marshall*, 1973].

Hueneme and Santa Cruz Canyons were included in our program
because they are similar in dimensions (Figures 3 and 4) but occupy
different sedimentologic settings. Hueneme Canyon incises the mainland shelf and is 10 km "downcoast" from the Santa Clara River, the
largest river in the southern California coastal watershed. Transparency measurements compiled by *Stevenson and Polaski* [1961] for
surface water and reconnaissance work by the present author indicated that the Hueneme Canyon area is characterized by relatively
high suspended sediment concentrations during all seasons. In contrast, the waters surrounding the northern Channel Islands are relatively clear [*Drake*, 1972] and sources of fine-grained terrigenous
sediment are minor. Since these two canyons are relatively close to
each other, are alike in overall size, yet can be considered "end-
members" in terms of suspensate concentrations, it was hoped that
the studies would provide a more conclusive test of the various
hypotheses of turbid-layer density flow.

SUSPENDED PARTICLES AND DENSITY STRATIFICATION

Early studies of the relationship between suspended particulate matter and water stratification in the sea suggested that
meaningful associations existed but, owing to the general lack of
continuous vertical profiles, the exact relationships were not resolved [*Jerlov*, 1959]. The recent widespread use of continuous,
high-resolution, salinity/temperature/depth (STD) systems has revealed the common presence of micro-stratification, adding a new
dimension to studies of oceanic water mixing [*Neal and Neshyba*,
1973].

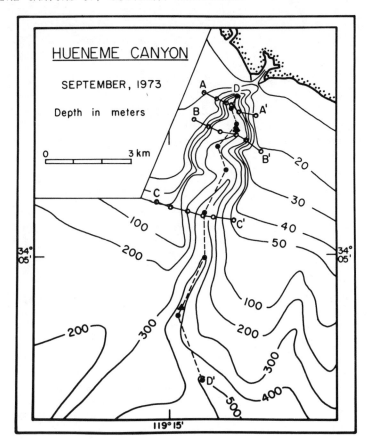

Fig. 3. Bathymetry and station locations for Hueneme Canyon.
Solid triangles show bottom current meter positions.

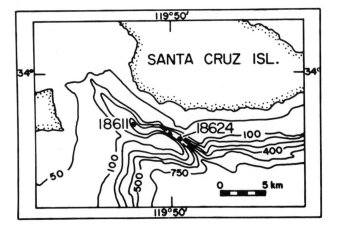

Fig. 4. Bathymetry and station locations for Santa Cruz Canyon.
Solid triangles show bottom current meter positions.

Continuous vertical light-transmission and temperature profiles
in the Santa Barbara Channel demonstrated a striking association
between steep segments of the thermal gradient and transparency reductions [*Drake*, 1971]. However, several vertical profiles revealed
steep thermal gradients having no detectable effect upon transparency and, conversely, strong turbid layers occupying nearly iso-
thermal water layers. In order to examine these relationships further, simultaneous light-transmission and STD recordings made in
submarine canyons off California are given in Figure 5. Inspection
of the representative profiles in this figure shows that the distribution of light-attenuating material is extremely sensitive to
changes in the density gradient; a change in this gradient results
in a change in the transparency gradient ranging from a slight slope
change to a complete reversal. Fifty records from four canyons were
analyzed to determine whether the turbidity peaks are associated
with the steepest portions of the pycnocline. The results show that
74% of such maxima fall at the top or in the center of steep pycnocline segments. The remaining turbidity maxima are centered within
the gentle portions of the density gradient (Figure 5) but, in
general, are relatively weak with < 5% transparency decrease. This
may mean that the light-attenuating material in these layers is
very fine grained or composed of low-density biogenic debris. For
this study, mid-water turbid layers were sampled on several occasions for microscopic study. While the few samples for which this
analysis is complete demonstrate that these layers are composed of
predominantly inorganic detritus, no meaningful statements on possible contrasts in the particles contained within "major" and "minor" maxima can be made. In the future, our research effort will
shift toward more intensive study of the characteristics of particles incorporated in turbidity maxima. At present, our data support the conclusion that turbid layers are typically present within the steep portions of the vertical water density gradient. However, it is important to note that these relationships are based,
in all cases, upon data from within 50 km of the coast and, principally, from within 20 km.

TRANSPORT OF SUSPENSATE INTO THE CANYONS

Sediments covering the relatively wide Ventura shelf and potentially available for entrapment within Hueneme Canyon are contributed by the Santa Clara River, 10 km northwest of the canyon
(Figure 6). An intensive study utilizing drift cards, reported by
Kolpack [1971], covering a one-year period (1969) indicated southeastward flow of surface water over the inner shelf. This flow is
driven by northwest and westerly winds which predominate during all
seasons. Currents at all depths over the outer portions of the
Ventura shelf are directed to the northwest as shown by light transmission patterns, drift-card dispersal paths, and direct measurement of near-bottom currents [*Drake et al.*, 1972]. Although suspended terrigenous sediment is spread well out onto the shelf during periods of high-river discharge, much of the fluvial contribu-

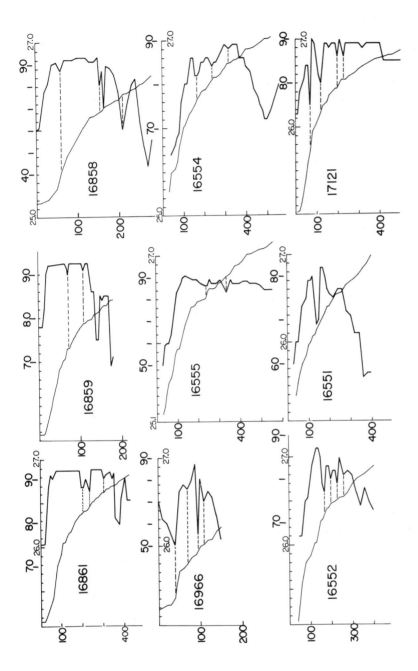

Fig. 5.   Representative vertical profiles of light beam transmission and sigma-t in southern California submarine canyons. Stations 16551 through 16555 are from Redondo Canyon; 16858 through 16861, Newport Canyon; and 17121, Hueneme Canyon. Depths are given in meters.

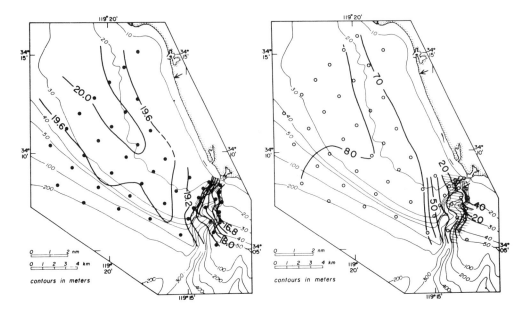

Fig. 6. Temperature distribution of the surface water over
the Ventura shelf and Hueneme
Canyon, September, 1972. The
Santa Clara River is indicated
by the arrow.

Fig. 7. Light transmission
(%T/m) of the surface water,
Hueneme Canyon area, September,
1972.

tion remains nearshore and is transported parallel to the coast toward Hueneme Canyon (Figure 7). Currents in the Hueneme area are
notoriously complex and variable [Gorsline, 1970] owing to the wind-
driven upwelling characteristic of the shelf east of the canyon and
to the confluence of westward regional currents and eastward-flow-
ing wind drift currents [Kolpack, 1971] near this coastal promontory.
During September 1972, transparency and temperature patterns of the
surface water clearly show a southeasterly drift of nearshore turbid water which flowed seaward along a marked convergence located
over Hueneme Canyon (Figure 7). One day later, beam transmission
values over the canyon had increased 30 - 50% at all stations,
demonstrating the rapid and large variations found to be a characteristic of suspended sediment concentrations within 5 km of the
coast off southern California.

Whereas local variations over shelf can be large, it can be concluded with reasonable confidence that high surface water concentrations of particulate matter are generally confined to the inner
shelf off southern California [Drake, 1972; Drake and Gorsline,
1973] in agreement with the work of Manheim et al. [1970; 1972]
along the Gulf coast and eastern continental margins of the United
States. Of greater importance to the movement of suspended sediment into Hueneme Canyon is the distribution of suspended particles
just above the sea floor. Water transparency values and filtered

water samples taken at 1 m above bottom in September 1972 (Figure
8) confirm the presence of a nepheloid layer containing approximately 0.6 mg/ℓ of suspended particles at the shelf edge (100 m) and
more than 6.0 mg/ℓ over the inner shelf. Particle concentrations
along the canyon axis exceeded 0.6 mg/ℓ to a depth of about 450 m,
with a peak concentration of 4.6 mg/ℓ at an axial depth of 30 m.
The tongue-like seaward extension of turbid water along the canyon
floor implies a decided downcanyon transport of suspended particles
just above the sea bed.

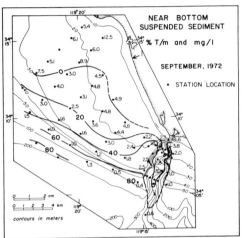
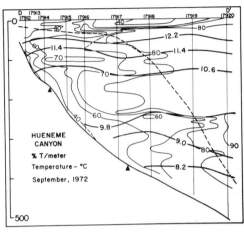

Fig. 8. Light-transmission
values and particle concentrations
at 1 m above the sea floor,
Hueneme Canyon area, September
1972. The particle concentrations
were determined by Millipore filtration of samples obtained using
a 30-ℓ Niskin bottle.

Fig. 9. Axial profile of
light beam transmission and
temperature in Hueneme Canyon,
September 1972. Solid triangles
show the positions of bottom
current meters. Depths are
given in meters.

PARTICLE DISTRIBUTIONS WITHIN THE CANYONS

     Suspended particle concentrations in all southern California
canyons are highest within the surface mixed layer and within a
sometimes equally well-mixed bottom nepheloid layer. Both layers
become increasingly turbid as the coast is approached, with peak
concentrations ranging from approximately 1.0 to 4 mg/ℓ at the surface and 1.0 to 6.0 mg/ℓ just above the canyon floors [*Drake and
Gorsline*, 1973]. Concentrations at the surface typically decline
rapidly to values of 0.1 to 0.5 mg/ℓ at distances of 3-6 km from
shore, whereas near-bottom values remain relatively high (0.4-0.7
mg/ℓ) to the canyon mouths.
     Particle distributions at intermediate depths in mainland shelf

canyons are characterized by more or less sharply defined turbidity
maxima associated with the steeper portions of the water-density
gradient (Figure 5 and 9). The mid-water maxima become more turbid toward the heads of the canyons and, in those profiles where
stations are close together, it appears clear that the maxima are
rooted in the nepheloid layer (Figure 9 - 11). Seaward movement
of the suspended matter within the mid-water plumes is implied by
the gradual increase in transparency to the juncture of the canyon
walls, with the regional slopes followed by rapid dispersal of the
suspensate beyond the confines of the canyons (Figure 10a). Although it is apparent that particulate matter entering the canyon
heads escapes at intermediate depths as well as near the bottom,
only the currents within a few tens of meters of the canyon floors
have been studied in detail.

Our initial investigation of the distribution of particles in
the bottom nepheloid layer of Redondo canyon revealed a zone of
very turbid water in the central canyon between the depths of 250
and 400 m (Figure 11). A second survey several months later showed
the same secondary maxima, leading to a belief that it resulted
from the morphology of the canyon. However, subsequent work in
Newport and Hueneme Canyons has revealed similarly turbid zones
between 250 and 400 m [*Drake and Gorsline*, 1973]. Table 2 is a
summary of statistics of currents at the canyon floor as reported
by *Shepard and Marshall* [1973]. In Pacific Coast canyons, the near-
bottom currents oscillate up- and downcanyon with periods ranging
from 8-10 hours over the canyon fans to less than 2 hours in the
shallow canyon heads. *Shepard and Marshall* [1973] concluded that
the canyon-floor currents are related to tides in the lower portions
of the canyons and to internal waves at shallower depths. Of particular interest is the relatively sudden change in the period of

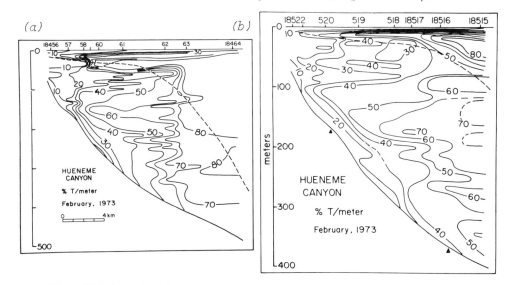

Fig. 10(a). Axial profile of light beam transmission measured
in Hueneme Canyon on February 13, 1973; and (b) February 15, 1973.

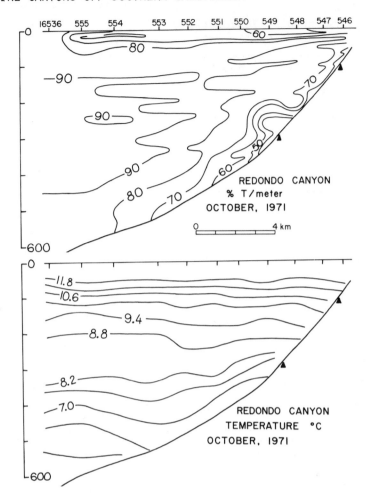

Fig. 11. Axial profiles of light transmission and temperature in Redondo Canyon in October, 1971. Solid triangles show the positions of bottom current meters. Depths are given in meters.

current reversal which apparently occurs in southern California canyons between 200 m and 350 m (Table 2). Reversal periods below about 250 m are dominated by tidal forces of longer periods, while those measured above 250 m are increasingly related to internal waves moving along the steeper pycnocline (Figure 3). *Emery* [1960] has shown that the water at intermediate depths (200-400m) over the southern California borderland is a mixture of southward flowing surface water of the California current and a deeper, high-salinity countercurrent from the south. An equal mix of these two water types generally occurs between 200 and 400 m, near the base of the relatively steep pycnocline. It is likely that the reversal periods of the canyon currents are influenced by this boundary.

The significant change in reversal period between the deep and

TABLE 2.   Depths, Highest Velocities and Flow Times of Submarine
              Canyons off Southern California*

Canyon	Depth(m)	Highest Velocity (cm/sec)	Average Flow Time (minutes)	
			Upcanyon	Downcanyon
La Jolla	46	26	65.6	53.5
La Jolla	78	17	76.0	60.0
La Jolla	167	29	75.0	94.0
La Jolla	206	29	74.0	97.0
La Jolla	375	22	216.0	247.0
Newport	101	17	87.0	105.0
Newport	252	11.5	168.0	255.0
Redondo	92	27	87.5	114.9
Redondo	283	19	231.0	91.0
Hueneme	173	27	144.0	126.0
Hueneme	373	38	192.0	104.0
Hueneme	445	34	306.0	270.0
Santa Cruz	357	36	201.5	279.0
Santa Cruz	582	29	126.0	372.0

* Modified from *Shepard and Marshall* [1973] and from *Shepard* (personal communication, 1973).

shallow portions of southern California canyons should lead to frequent periods of opposed near-bottom flows. It follows that particle maxima should be present wherever opposing currents converge. This explanation for the mid-canyon turbidity maxima is supported by axial profiles of light transmission and temperature in Redondo and Hueneme Canyons (Figures 9 - 11). These profiles were completed when near-bottom flow was predominantly upcanyon below 250 m and downcanyon above 250 m.

If opposing canyon-floor currents are a regular occurrence, they offer a convenient and reasonable explanation for the development of mid-water turbid layers in each canyon (Figures 9 and 11). Although detailed current meter measurements throughout the water volume of the canyon are needed to confirm this interpretation, it is most probable that these layers are produced by the seaward flow necessitated by convergences of canyon-floor currents.

TURBID-LAYER DENSITY FLOW

Several authors have discussed the possibility that suspended particulate matter incorporated in nepheloid layers may increase the water density sufficiently to produce "low-density" turbidity currents [*Bagnold*, 1963; *Postma*, 1968; *Moore*, 1969; *McCave*, 1972]. *Moore* [1969] believed that such a mechanism was required to explain the distribution of fine-grained desposits in the basins of the

California borderland. *McCave* [1972] developed the argument that
rates of sediment accumulation on the continental slopes and rises
of the world require some means of accelerated "low-level escape"
(i.e., near the sea bed) of silt and clay. However, as noted by
McCave, the low concentrations present over outer shelf areas preclude significant density underflow except in channelized situations,
such as in submarine canyons.

   Our work has demonstrated that nepheloid layers can be considered to be a permanent component of the suspensate distribution
in those canyons incising the mainland shelf off southern Californ
[*Drake and Gorsline*, 1973]. Furthermore, *Shepard and Marshall*
[1973] show that net downcanyon flow is also characteristic of the
near-bottom water in Pacific coast canyons. *Drake and Gorsline*
[1973] discussed the possibility that this permanent tubid layer
could influence the flow patterns in the Canyon, but concluded that
particle concentrations were far too low to initiate or maintain
through-going density-excess underflows. However, computation of
mean density-excess flow velocities using the known concentrations
and canyon slopes suggest that this transport mechanism could explain the net downcanyon flow of 1 to 2 cm/sec measured by Shepard.

   To test this important hypothesis further, a survey of the
bottom currents and suspended sediment distribution in Santa Cruz
Canyon was made (Figure 4). It was expected that suspended sediment concentrations in this offshore area would be low owing to the
limited sources of fine sediment and the coarse-grained shelly sands
which cover the insular shelf [*Booth*, 1973]. Although rough weather
and equipment breakdowns impeded a complete sampling program of suspended sediments, several vertical profiles of light transmission
were obtained along the canyon axis, near-bottom samples were recovered for gravimetric analyses, and currents were measured for 80
hours at depths of 357 and 582 meters. At the 357-m station, two
current meters were placed on one line at 3.6 m and 29 m above the
canyon floor. In Table 2 are shown the results of the current
measurements in both Santa Cruz and Hueneme Canyons in February
1973 (F. P. Shepard, unpublished data). The bottom currents in
Santa Cruz Canyon flow predominantly up- and downcanyon, although
weak cross-canyon components were appreciable at the shallower of
the two stations. The marked difference between the total durations
of the up- and downcanyon flows should be noted. While average current velocities were nearly the same in both directions, downcanyon
flows at 582 m lasted more than twice as long as the upcanyon currents. In effect, net transport of bottom water was decidedly downcanyon at both stations, ranging from ~65 m/hr at 357 m to 155 m/hr
at 582 m; the latter value is among the strongest net downcanyon
flows measured by Shepard.

   Figure 12 shows two of eight vertical light-transmission and
temperature profiles obtained in this canyon during the current-
recording period. Below the surface mixed-layer, transparency
variations were minor even though several abrupt thermal discon-
tinuities were present. In particular, decreases in light transmission near the canyon floor never exceeded 4% and typically were
only 1-3%. Samples of water taken 2 m above the canyon floor using

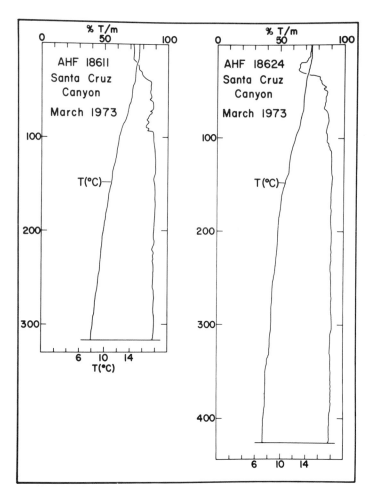

Fig. 12. Profiles of light beam transmission and temperature, Santa Cruz Canyon. Refer to Fig. 4 for station locations.

a bottom-activated Niskin bottle contained 120 µg/ℓ of suspensate at station 18611 and 105 µg/ℓ at station 18624. Samples taken at 100 m above the canyon floors contained 92 µg/ℓ and 115 µg/ℓ at these stations. It is clear that the near-bottom turbidity increases resolved by the transmissometer were so slight as to be below the detection limits of the filtration procedures. Figure 2 indicates that the maximum particle increase near the bottom could not have exceeded approximately 20 µg/ℓ. The influence which this amount of suspended detritus might have upon the flow of bottom water can be estimated using the equation of *Middleton* [1966]:

$$\bar{U} = [8g\Delta p/pRS/f]^{\frac{1}{2}}$$

where $\bar{U}$ is the mean velocity, $\Delta p/p$ is the density-excess due to the suspensate, R is the hydraulic radius of the canyon, S is the axial

slope and f is the total bottom and upper interface friction.
Values for all terms in the equation, except f, are easily specified.
A value of about 0.03 for f is probably reasonable, based on the
experiments of *Middleton* [1966]. Using 2×10^{-8} for Δp, 20 m for
R and 0.07 for S, the maximum current which could be developed by
the suspended matter is about 0.5 cm/sec. Thus, the low concentrations of the nepheloid layer in Santa Cruz Canyon can, under the
most favorable cirucmstances, account for only ~30% of the net
flow at 357 m and <10% at 582 m.

The current records for Hueneme Canyon are unusual in that they
show stronger average upcanyon velocities and nearly equal durations
for up- and downcanyon flows. In addition, the peak speeds at the
two deep stations are the highest (48cm/sec) ever recorded by
Shepard in submarine canyons, whereas peak and mean speeds were
lower at the shallow station. A complete discussion of these data
will be presented by F. P. Shepard elsewhere. In any case, it is
significant that in spite of the high concentrations of suspended
particles in the Hueneme Canyon, net flow was upcanyon at 172 m
(~0.3 cm/sec) and weakly downcanyon at the deeper stations. These
data, combined with the data from Santa Cruz Canyon, strongly suggest that the concentrations of suspended solids normally present
in canyons off southern California are too low to exert a measureable
influence on the current patterns of a canyon.

Although the low-density, turbid-layer underflow mechanism within canyons has a reasonable physical basis, it is evident that suspended particulate matter is not required to produce the downcanyon
drift of canyon bottom water. At present, I believe that the current regime of a canyon is closely related to the dynamics of internal gravity waves off southern California. In theory, progressive internal waves sweeping shoreward across the continental margin should generate seaward transport of bottom water [*Lafond*, 1962]
owing to the relatively vigorous seaward flow below wave troughs.
Furthermore, while the decreasing cross-sectional area of canyons
should intensify wave-generated flows, currents related to internal
waves should also be important on open slopes and plane shelves.
The break in the mainland shelf off southern California ranges from
75 m to 100 m in depth [*Emery*, 1960] and, during most of the year,
coincides with the base of the steep pycnocline separating surface-
and intermediate-depth waters (Figure 5). Internal waves moving
along this density boundary would, therefore, expend much of their
energy on the outer shelf and upper slope. Unfortunately, investigations of internal waves deeper than about 60 m off southern California have not been undertaken [*Lafond*, 1962; *Summer and Emery*,
1963]. Indirect comments on the potential significance of sediment
movement by internal waves are common in literature but this aspect
of sedimentation of the continental margin has not received adequate
attention.

MOVEMENT OF SUSPENDED SEDIMENT THROUGH THE CANYONS

While considerably more data are needed on the current structure
throughout the canyon volume, it is possible to compute a crude esti-

mate of suspended sediment flux through southern California mainland shelf canyons. There are seven major canyons incising the
mainland shelf off southern California [Emery, 1960]. If it is
assumed that seaward flow of 1 cm/sec occurs within a near-bottom
layer that is 20 m thick and 500 m wide and contains 5 mg/ℓ of particles (this extreme value is selected to illustrate the point),
approximately 1.1 x 10^5 tons of suspended particles would move
through the seven canyons each year. Rodolfo [1970], Moore [1969]
and Emery [1960] have estimated the amount of sediment entering
the offshore region from the coastal watershed as ranging from 7 to
10 million metric tons annually with about 50% of this total being
silt and clay. Thus the flux through submarine canyons near the
bottom accounts for only about 2 to 3% of the total terrigenous suspended sediment contribution. Even assuming that an equivalent flux
of material occurs along 2 other layers in each canyon, the total
canyon transport would still be well below 10% of the annual watershed contribution. Clearly the bulk of the fine particulate matter
must either remain on the shelf, escape along pathways which do not
include the canyons, or move seaward as high-density turbidity currents.

CONCLUSIONS

Mid-water light-transmission reductions within submarine canyons off southern California are associated predominantly with the
steeper segments of the vertical water-density gradient. Advective
seaward flow from permanent nepheloid layers is probably an important process supplying particles to the mid-water turbid plumes.

The escape of suspended particles from canyons appears to occur
at several levels but is best documented within the nepheloid layer
where Pacific coast canyons have shown net downcanyon flow ranging
from 0 cm/sec to 3.4 cm/sec [Shepard and Marshall, 1973]. Estimates
of the annual flux of fine-grained material through all southern
California mainland shelf canyons indicates that less than 10-15% of
the annual terrigenous supply moves through the canyons.

Typical concentrations of suspended sediment over the floors of
mainland shelf canyons range from about 0.4 to 6 mg/ℓ. Owing to
increases in water density with depth, such concentrations are far
too low to initiate through-going turbid underflows [Drake and
Gorsline, 1973]. Earlier calculations of the influence the nepheloid layers might have upon canyon-floor currents suggest that the
slow, net downcanyon flow demonstrated by Shepard and Marshall
[1973] could be caused by the suspended matter. However, more recent data from Santa Cruz Canyon do not support this conclusion.

ACKNOWLEDGEMENTS

I wish to thank F. P. Shepard, N. Marshall, P. McLoughlin and
G. Sullivan of Scripps Institution of Oceanography for help at sea

and ashore; R. Loudermilk and the staff of the Visibility Laboratory, Scripps Institution of Oceanography, for technical advice; D. S. Gorsline and J. S. Booth, University of Southern California, for many fruitful discussions; and R. L. Kolpack of the University of Southern California for use of his current-meter data.

This research was carried out on board the R/V *Velero IV* of the University of Southern California and was supported by National Science Foundation grants GB-8206 and GA-22842.

REFERENCES

Bagnold, R. A., Mechanics of marine sedimentation, in *The Sea*, vol. 3, edited by M. N. Hill, pp. 507-528, Interscience, New York, 1968.

Booth, J. S., Textural changes as an indicator of sediment dispersion in the Northern Channel Island Passages, California, *J. Sed. Petrology*, *43*, 238-251, 1973.

Cook, D. O., and D. S. Gorsline, Field observations of sand transport by shoaling waves, *Mar. Geol.*, *13*, 31-55, 1972.

Drake, D. E., Suspended sediment and thermal stratification in Santa Barbara Channel, California, *Deep Sea Res.*, *18*, 763-769, 1971.

Drake, D. E., Suspended matter in Santa Barbara Channel, California, Ph.D. thesis, 357 p, University of Southern California, Los Angeles, 1972.

Drake, D. E., R. L. Kolpack, and P. J. Fischer, Sediment transport on the Santa Barbara-Oxnard shelf, Santa Barbara Channel, California, in *Shelf Sediment Transport: Process and Pattern*, edited by D. J. P. Swift, O. Pilkey, and D. Duane, pp. 307-331, Dowden, Hutchinson and Ross, Stroudsburg, Pa. 1972.

Drake, D. E., and D. S. Gorsline, Distribution and transport of suspended particulate matter in Hueneme, Redondo, Newport and La Jolla submarine canyons, *Bull. Geol. Soc. Amer.*, *84*, 3949-3968, 1973.

Emery, K. O., *The Sea Off Southern California*, J. Wiley and Sons, New York, 366 pp., 1960.

Gorsline, D. S., Reconnaissance survey of the hydrographic characteristics of the Hueneme-Mugu shelf, June-July, 1970, *Rept. USC Geol. 70-6*, Univ. of Southern Calif., Los Angeles, 1970.

Gorsline, D. S., Annual report on marine geologic research in the California Continental Borderland: Heavy mineral content of river, beach and nearshore sands, southern California, *Rept. USC Geol. 71-5*, Univ. of Southern Calif., Los Angeles, 1971.

Ingle, J. C., *The Movement of Beach Sand*, Developments in sedimentology, vol. 5, Elsevier, New York, 1966.

Inman, D. L., and T. K. Chamberlain, Littoral sand budget along the southern California coast, *Report of 21st Intl. Geol. Cong.*, Copenhagen, pp. 245-246, 1960.

Jerlov, N. G., Maxima in the vertical distribution of particles in the sea, *Deep Sea Res.*, *5*, 173-184, 1959.

Jerlov, N. G., *Optical Oceanography*, Elsevier, Amsterdam, 194 pp., 1968.

Kolpack, R. L., (editor), *Biological and Oceanographical Survey of the Santa Barbara Channel Oil Spill*, vol. II: Physical, Chemical and Geological Studies, 477 pp., Allan Hancock Foundation, Univ. Southern California, Los Angeles, California, 1971.

Lafond, E. C., Internal waves, in *The Sea*, vol. 1, edited by M. N. Hill, pp. 731-751, Interscience, New York, 1962.

Manheim, F. T., R. H. Meade, and G. C. Bond, Suspended matter in surface waters of the Atlantic continental margin from Cape Cod to the Florida Keys, *Science*, *167*, 371-376, 1970.

Manheim, F. T., J. C. Hathaway, and E. Uchupi, Suspended matter in surface waters of the northern Gulf of Mexico, *Limnol. Oceanogr.*, *17*, 17-27, 1972.

McCave, I. N., Transport and escape of fine-grained sediment from shelf areas, in *Shelf Sediment Transport: Process and Pattern*, edited by D. J. P. Swift, O. Pilkey, and D. Duane, pp. 225-248, Dowden, Hutchinson and Ross, Stroudsburg, Pa., 1972

Middleton, G. V., Experiments on density and turbidity currents, *Can. Jour. Earth Sci.*, *3*, 627-637, 1966.

Moore, D. G., Reflection profiling studies of the California Continental Borderland: Structure and Quaternary turbidite basins, *Geol. Soc. Amer. Spec. Paper 107*, 142 pp., 1969.

Neal, V. T., and S. Neshyba, Microstructure anomalies in the Arctic Ocean, *J. Geophys. Res.*, *78*, 2695-2701, 1973.

Oguri, M., and R. Kantor, Primary productivity in the Santa Barbara Channel, in *Biological and Oceanographical Survey of the Santa Barbara Channel Oil Spill*, edited by D. Straughan, Univ. So. Calif., Los Angeles, 1971.

Postma, H., Suspended matter in the marine environment, in *Some Problems of Oceanology*, Intl. Oceanographic Cong. 1966, State Publ. House, Moscow, pp. 258-265, 1968.

Petzold, T., and R. Austin, An underwater transmissometer for ocean survey work, *Scripps Inst. Oceanogr. Tech. Report 68-9*, 5 pp., 1968.

Rodolfo, K. S., Annual suspended sediment supplied to the California Continental Borderland by the southern California watershed, *J. Sed. Petrology*, *40*, 666-671, 1970.

Shepard, F. P., and R. F. Dill, *Submarine Canyons and other Sea Valleys*, Rand McNally, New York, 381 pp., 1966.

Shepard, F. P., and N. F. Marshall, Currents along the floors of submarine canyons, *Amer. Assoc. Petrol. Geologists*, *57*, 244-264, 1973.

Stevenson, R. E., R. B. Tibby, and D. S. Gorsline, The oceanography of Santa Monica Bay, California, Rept. to Hyperion Engineers, Geology Dept., Univ. So. Calif., Los Angeles, 268 pp., 1956.

Stevenson, R. E., and W. Polski, Water transparency of the southern California Shelf, *Calif. Acad. Sci. 60*, 77-87, 1961.

Summers, H. J., and K. O. Emery, Internal waves of tidal period off Southern California, *J. Geophys. Res.*, *68*, 827-839, 1963.

Terry, R. D., S. A. Keesling, and E. Uchupi, Submarine geology of Santa Monica Bay, California, Rept. to Hyperion Engineers, Inc., Geology Dept., Univ. So. California, 177 pp., 1956.

Trask, P. D., Movement of beach sand around southern California

promonitories, Dept. of Army, Corps of Engineers, Beach Erosion
Board, *Tech. Memo. 28*, 24 pp., 1955.
Vernon, J. W., Shelf sediment transport system, Ph. D. thesis,
University of Southern California, 135 pp., 1966.

Continuous Light-Scattering Profiles and Suspended Matter Over Nitinat Deep-Sea Fan

EDWARD T. BAKER, RICHARD W. STERNBERG, AND
DEAN A. McMANUS

University of Washington

ABSTRACT

During September 1971 and June 1972, a total of 65 light-scattering profiles were recorded in the waters over Nitinat Fan by means of a self-contained, continuously recording nephelometer. All profiles extend from the sea surface to a point 20 m above the sea floor (maximum depth 2400 m) and as many as 5 prominent scattering layers are observed within the water column at a given station. Some scattering layers can be traced over wide areas of the fan and appear on records from both cruises. The most persistent feature of the profiles, and the only one found on every record, is a bottom nepheloid layer (BNL) of steadily increasing light scattering immediately below a layer of relatively clearer water and immediately above the sea floor. Over short lateral distances the vertical extent and scattering intensity of the BNL (normalized to the scattering levels of the overlying clearer waters) change markedly in a manner apparently related to fan topography. In general, the BNL thickens to > 300 m and intensifies to > 2.5 times normal over the topographic lows, being most prominent above Cascadia Valley (the major valley crossing the fan) as well as above the steep non-channelled northern flank of the fan. Over the levees separating the smooth northern flank from Cascadia Valley to the south and over the foot of the continental slope which forms the eastern border of the fan, the BNL thins to < 50 m and the scattering intensity is < 1.5 times normal.

Twenty-four suspended sediment samples, collected by in situ filtration from within and just above the BNL, yielded inorganic particle concentrations of 20 to 80 µg/ℓ. These concentrations correlate well (r = 0.73) with the simultaneously recorded

155

*scattering values, indicating that the nephelometer provides a
fair estimate of suspended sediment concentrations in the bottom
waters over Nitinat Fan.*

*By integrating the scattering profile within the BNL at each
station and converting the average scattering value into particle
mass by means of an empirical relationship derived from the measured
concentration values, the average amount of inorganic suspended
matter in the BNL can be calculated. In a 1-cm^2 column of water,
average particulate mass in the BNL was estimated at about 0.5 mg.
Assuming that the bottom deposits are derived only from particles
in the BNL, the measured sedimentation rate on the fan requires a
mean residence time of about 1 month for these particles within the
BNL.*

INTRODUCTION

During the past half-decade, *Thorndike and Ewing* [1967] have
used a photographic nephelometer to catalogue the principal
light-scattering features of many of the deep ocean basins.
Records have been collected in the Arctic Ocean [*Hunkins et al.,*
1969], the Atlantic [*Eittreim and Ewing,* 1972; *Ewing et al.,* 1971],
the North Pacific [*Ewing and Connary,* 1970], and Antarctic [*Eittreim
et al.,* 1972] and elsewhere. None of these studies, however, has
examined the temporal and spatial distribution of light-scattering
particles in bottom waters with regard to a particular deep-sea
bathymetric feature. The focus of the present investigation has
consequently been directed toward that topic. Continuous light-
scattering profiles and discrete *in situ* suspended sediment
samples were employed to record the nature and variability of the
light-scattering particles over a prominent deep-sea sedimentary
feature on a scale sufficiently small for recognition of any
correlation of the suspensate patterns with the sea floor topography
and sedimentation history.

The study area chosen for the investigation is Nitinat Fan which
lies seaward of Nitinat and Juan de Fuca submarine canyons and at
the foot of the Washington continental slope (Figure 1). The
re-entrant in the 1800-m contour in the upper right-hand portion of
the figure is the combined mouth of the Juan de Fuca and Nitinat
submarine canyons. Cascadia Valley, the major topographic feature
of the fan, extends westward from these canyons then curves abruptly
to the southeast and widens considerably, exhibiting what appears
to be a braided-stream type of topography. Farther south, the
channel narrows to a prominent V-shape which continues southward.
Seaward, the valley is bordered by pronounced natural levees. To
the northwest, the topography of the fan is a smooth, somewhat
steeply dipping slope with no connection to the canyons that egress
on the apex of the fan. To the southwest, the topography is complex,
showing remnants of many smaller subsidiary valleys.

A total of 65 nephelometer profiles, each extending from the
sea surface to 20 m off the sea bottom, were recorded during two
research cruises over Nitinat Fan: cruises TT63 and TT68 of the

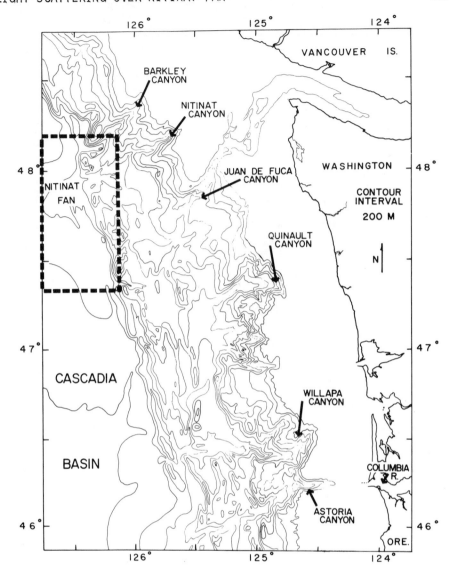

Fig. 1. Location map of the study area. Nitinat Fan is situated at the foot of the Washington continental slope.

R/V *Thomas G. Thompson* in September 1971 and May 1972 respectively (Figure 2). In addition, 24 samples of near-bottom suspended sediments were collected during cruise TT68 at 18 of the stations over the fan using an *in situ* filter-pump system.

INSTRUMENTATION METHODS

The design of the integrating nephelometer used in the present

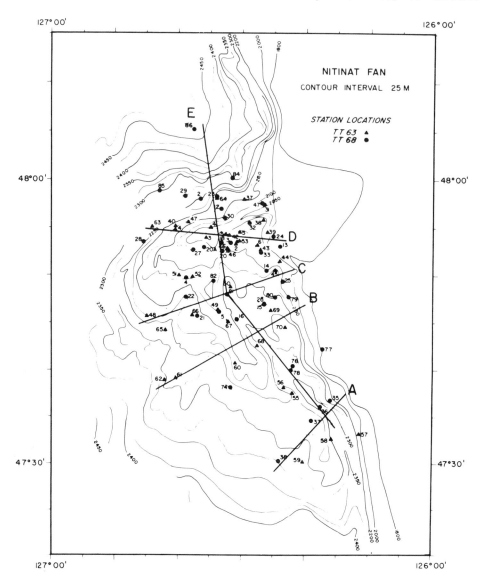

Fig. 2. Nephelometer stations over Nitinat Fan. Profile lines
A-E mark the location of the cross sections for the bottom
nepheloid layer (BNL) shown in Figure 10.

study is based on a principle introduced by *Beutell and Brewer*
[1949]. Its operational components closely resemble those of an
atmospheric nephelometer designed at the Atmospheric Chemistry
Laboratory at the University of Washington [*Ahlquist and Charlson*,
1968, 1969; *Charlson et al.*, 1969]. The geometry of the scatterance
meter of Beutell and Brewer and of the instrument used in this study
is shown in Figure 3.

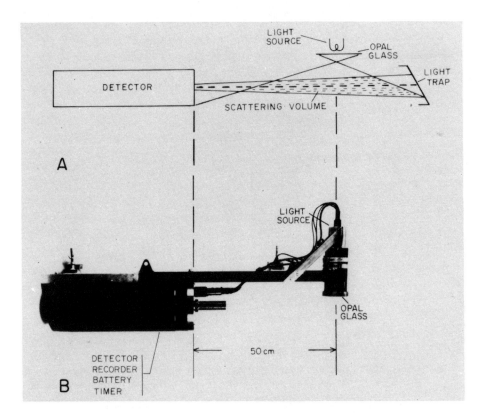

Fig. 3. (*a*) Geometry of the scatterance meter of *Beutell and Brewer* [1949]. (*b*) The nephelometer used in this study.

The nephelometer consists of a flashing light source, a scattered-light detector system, a strip-chart recorder, and the necessary battery power mounted in a self-contained portable housing. The instrument continuously measures the total scattering of light from suspended particulate matter and can be lowered to depths of 4000 m. The scattering profiles were recorded internally on a strip-chart recorder, allowing the data to be observed directly and utilized immediately upon completion of a profile. A complete description of the instrument system is given by *Sternberg et al.* [1974].

The *in situ* filter-pump system is shown in Figure 4. It was designed as a large-volume sampler for collection of sufficient amounts of particulate matter for mineralogic analysis by X-ray diffraction. In the deep sea, the filter-pump system normally filters approximately 400 ℓ of seawater during a 30-min pumping cycle. Seawater is drawn through a 293-mm membrane filter having a mean pore size of 0.45μ and through a cumulative flowmeter by a

Fig. 4. The *in situ* filter-pump system. Components shown are: (*a*) stainless steel membrane filter holder, (*b*) flowmeter readout, (*c*) flowmeter, (*d*) pump motor in pressure-compensated case, (*e*) timing system, (*f*) lead-acid batteries, (*g*) exhaust tube (extends 7m above the filter intake during use).

flexible impeller pump. It is then pumped through a tube and expelled at a point about 7 m above the filter intake. Power is provided by lead-acid batteries completely encased in a pressure-compensating container. The unit is programmed at the surface by two manually adjustable time switches, the first of which is set to turn the pump on at a preset time after the unit is lowered to the appropriate depth; the second switch is set to turn the pump off after a given pumping time of up to one hour.

RESULTS

In order to convert the light-scattering data into a geologically useful format, it is necessary to calibrate the nephelometer output in terms of some physical property of the particles, such as mass concentration. Absolute quantification of the light scattering data is a complicated and necessarily inexact procedure, since the total scattering value of a parcel of water is a complex integral of all the properties of all the particles, of which particle mass concentration is only one, albeit the major, factor [*Jerlov*, 1968]. Thus, the most appropriate and simplest method of calibration appear-

ed to be use of an empirical technique for comparing the recorded
light-scattering values to simultaneously collected values of particle mass concentration. Using the *in situ* filter pump, 24 samples of suspended sediment were collected from the bottom waters at
18 stations over Nitinat Fan. Concentration values of inorganic
suspended sediment — i.e., the material not oxidizable by H_2O_2 —
typically range between 20 and 80 µg/ℓ and show a marked correlation
with values of light scattering. The resulting regression curve,
shown in Figure 5, is given as:

$$CONC \text{ (µg/ℓ)} = 6.0 \text{ (scattering)} + 13.0$$

The sampling program yielded concentration values of only the
inorganic — and, therefore, geologically important — fraction of
the suspended matter; a better fit may be obtained by plotting total
suspended load against light scattering. Furthermore, the data as
yet obtained cover only a limited geographical area and concentration range, and an assumption that the correlation is linear for all
regions and concentrations may be incorrect. Waters with very di-
lute (<15 µg/ℓ) or very concentrated (>500 µg/ℓ) suspensions may
show correlations which differ appreciably from the trend presented,

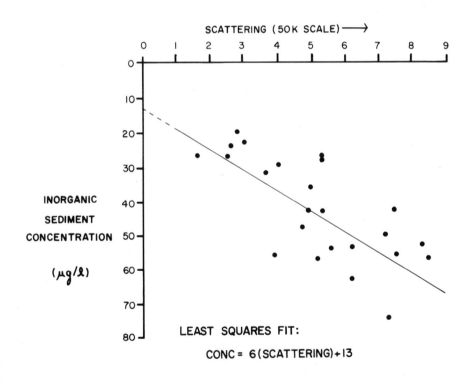

Fig. 5. Concentration of inorganic suspended sediment versus light
scattering for samples from cruise TT68 within the BNL and the overlying clearer water. Note that the curve is not defined for concentration values below about 20 µg/ℓ or above about 80 µg/ℓ.

due to differences in the particle-size distribution, organic/
inorganic suspensate ratios, and other parameters of natural
suspensions.

Figure 6 shows 6 representative nephelometer profiles from

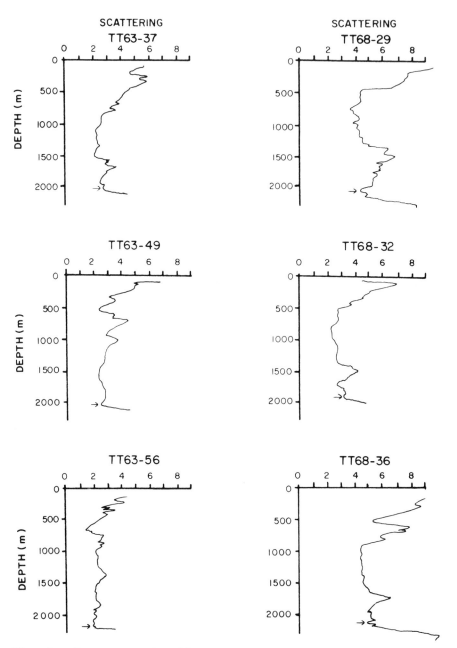

Fig. 6.   Representative light-scattering profiles over Nitinat Fan.
Note the many prominent mid-depth scattering maxima.   Small arrows
mark the top of the BNL.

cruises TT63 and TT68 which illustrate the range of light-scattering
features found over Nitinat Fan. Characteristic features of these
profiles include high scattering in the surface waters, less intense
scattering in the mid-depth regions with numerous individual maxima,
and a relatively thin bottom layer of high scattering directly
above the sea floor.
     Although the nephelometer profiles in Figure 6 show many
individual features they can be grouped on the basis of certain
striking similarities summarized by the idealized profiles in
Figure 7. It is noteworthy that the most common scattering maxima

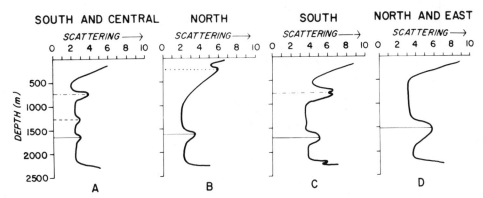

Fig. 7. Idealized nephelometer profiles, showing the depths of
the most common scattering maxima found over Nitinat Fan.

and minima are not found randomly throughout the area. Rather, during both cruises, conducted at different seasons of the year, the
profiles appear to be divided according to approximately northern
and southern geographical units, with those in the central part of
the study area often not fitting well into either classification.
Furthermore, there are large-scale features which are common to
more than one area and time of the year. The thick scattering layer
centered between 1500-1800 m — an apparently non-seasonal,
widespread scattering phenomenon — is one of the most common
features noted on the profiles. The upper 1000 m of the water
column is characterized by one of three basic scattering configurations: (1) a steady decrease in scattering to 400-500 m, a high
scattering layer at about 750 m, and another minimum at 1000 m
(Figure 7*a* and *c*); (2) a steady decrease through the first 500 m
with little change in the second 500 m (Figure 7*d*); or (3) a
continual decrease to 1000 m, often with a thin scattering layer
superimposed at 300-400 m (Figure 7*b*).
     The most persistent and geologically interesting feature of the
profiles is the bottom nepheloid layer (BNL), the zone of increasing
light scattering immediately overlying the sea floor and within
which there are no scattering levels lower than that found at the
top of the zone. The BNL appears on all 65 profiles recorded over
Nitinat Fan, just as it is nearly ubiquitous in profiles from most

of the deep-ocean basins.

The distribution of thickness and scattering intensity of the local BNL is shown in Figures 8 and 9, composite maps of data compiled from both cruises. It seems particularly significant

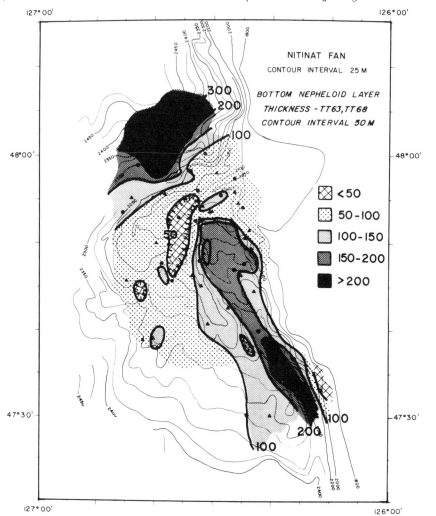

Fig. 8. Thickness of the BNL over Nitinat Fan. Thicknesses are greatest over the northwest flank of the fan and over Cascadia Valley. The BNL thins substantially over the levee separating these areas.

that composite maps can be drawn, suggesting that the primary characteristics of the BNL remained relatively stable throughout the 9-month interval between the fall and spring cruises.

The greatest BNL thickness (> 300 m) was encountered in the deep waters over the smooth northern flank of the fan, an area sampled only during cruise TT68. Toward the south, the BNL thins substantially as it crests the seaward levee of Cascadia Valley.

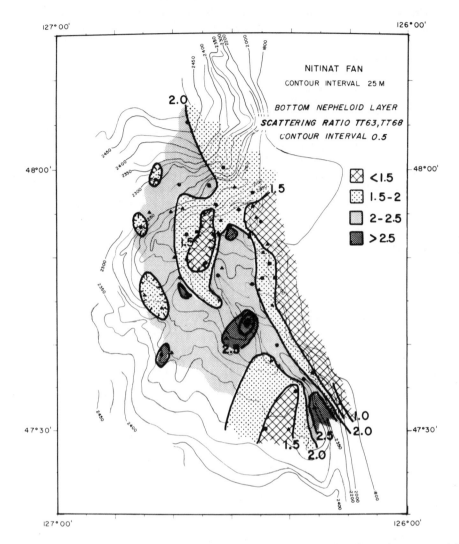

Fig. 9.   Scattering intensity of the BNL, calculated as a ratio
of the scattering value at the bottom of the BNL to the scattering
value of the clear water just above the BNL.   The scattering ratio
exhibits a positive correlation with the thickness of the BNL.

The thinnest BNL (< 50 m) of the entire area occurs over the levee
near the apex of the fan where Cascadia Valley changes from a
southwest to a southeasterly direction.   This configuration was
prominent during both cruises TT63 and TT68.
     At the head of Cascadia Valley, BNL thicknesses average about
75-100 m increasing to 100-200 m as the valley widens and turns to
the southeast.   At station TT63-50, a 200-m thick layer over the
western side of the valley just at the foot of the levee was
recorded — a fourfold increase over the 50-m thick layer found just

6 km to the west. Farther south, where Cascadia Valley narrows and assumes a prominent V-shape cross section, BNL thicknesses increase to 200 m.

As the continental slope rises sharply along the eastern side of Cascadia Valley, the nepheloid layer thins to < 100 m a short distance up the slope. Thicknesses to the west of Cascadia Valley, seaward of the levee, are less coherent, ranging between 30 and 150 m. This may be a function of the complex nature of the topography found over the fan's southwestern corner, or, perhaps, a result of the limited sample coverage. Only seven stations represent this large area.

The scattering intensity of the BNL (Figure 9), calculated as a ratio of the scattering value at the bottom of the BNL to the scattering value of the clear water just above the BNL, shows a positive correlation with the BNL thickness. The scattering ratio exceeds 2.0 over the northwest flank of the fan and over the main body of Cascadia Valley. Ratios are lowest over the valley levee near the fan apex and along the base of the bordering continental slope. Over the southwestern region of the fan the scattering ratios, as found with the BNL thickness, form no coherent pattern.

Cross sections drawn through the BNL (Figure 10), with super-imposed intensity gradients, add a vertical perspective to the areal maps presented above. These figures were constructed by choosing 5 representative cross-sectional areas (refer to Figure 2 for positions of the cross sections) and then projecting onto each line all nearby stations from both cruises. An "average" BNL surface was drawn through these stations, as were the contours indicating the relative intensity within the BNL.   The lowermost 20 m is not contoured due to the absence of data in that zone.

It is observed on these cross sections, particularly the north-south oriented cross section E, that the relief on the BNL surface has the effect of "softening" the underlying topography of Nitinat Fan itself. Not only the BNL surface, but the scattering intensity contours "on lap" the crest of the levee and the foot of the continental slope. The effect is most striking near the central part of the fan on cross sections C, D, and E in Figure 10. Farther south, as the valley/levee relief lessens, the influence of the levee topography on the horizontal continuity of the BNL intensity gradients is less pronounced (Figure 10, cross sections A and B).

DISCUSSION

The complementary patterns exhibited by the BNL thickness and intensity parameters create a relatively consistent picture of the nature of the turbidity of the bottom waters over Nitinat Fan. In general, the BNL tends to fill in the topographic lows (the northwestern flank and Cascadia Valley) leaving the levees and the marginal ridge at the base of the continental slope protuding through the more intense portions of the BNL.

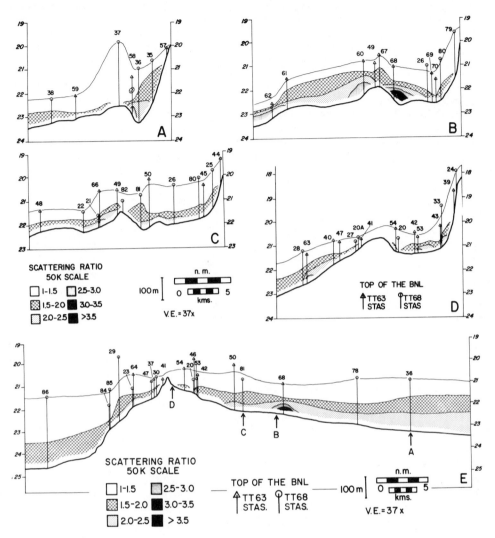

Fig. 10. Cross sections of the BNL with superimposed scattering-ratio gradients. Topographic influence on the BNL continuity is particularly apparent in sections *C*, *D*, and *E*. Refer to Figure 2 for location of the cross sections.

Above the BNL, vertical fluctuations in the light-scattering intensity may be related to the water mass structure and currents found in the area. Although the detailed hydrography of this region is not well known, *Ingraham* [1967] has shown that relatively colder and less saline Pacific Subarctic Water appears to be moving on-shore from the west between 400 and 1000 m, coinciding with the depth of the regional zone of minimum oxygen (300-900 m). Below and above this depth interval the waters are more characteristic of the northward flowing Pacific Equatorial Water Mass [*Ingraham*, 1967].

Comparison of these observations of the Subarctic Water Mass and the oxygen-minimum with the light-scattering curves offers several interesting speculations. For most of the profiles, major changes in light scattering occur at approximately 500 and/or 1000 m. Also, many profiles, particularly those over the southern half of the fan, show a high-intensity scattering layer in the center of this zone at about 750 m. Whether these and other features can be related directly or indirectly to water mass properties and density structure awaits further synoptic light-scattering and hydrographic field measurements, but the apparent permanence and extent of many of the scattering layers suggests at least some correlation with the non-seasonal water structure.

The distribution of light scattering within and above the BNL, together with the nephelometer calibration (Figure 5), provide a means to estimate the inorganic suspended sediment concentration in the waters over Nitinat Fan. The concentration measurements, along with prior calculations of the sedimentation rate on the fan [*Carson*, 1971] allow a first estimate of the sediment budget over Nitinat Fan.

Integrating the scattering profile at each station and converting the average scattering value into particle mass yields the average mass of inorganic suspended matter in the water column. Figure 11 shows the resulting values for a 1-cm^2 column of water, both within and above the BNL, averaged for each cruise over Nitinat Fan. Comparable values are also given for a station at the mouth of Juan de Fuca Canyon, which exits from the slope at the head of Cascadia Valley. (The values do not include the uppermost 200 m of the water column, because of errors due to interference from high concentrations of living plankton.)

The differences observed in the data from the fall and spring cruises may be partially seasonal, reflecting a possible change in sediment supply due to the shift from ocean storms and high continental runoff in the spring to quiet waters and minimum runoff in the early fall. The evidence is as yet insufficient to draw a firm conclusion, particularly without knowledge of the time lag between the introduction of sediment into the coastal waters and its appearance over Nitinat Fan.

Apart from the effect of seasonal fluctuations, an average value for suspended inorganic sediment within the BNL is about 0.45 mg/cm^2; the water column above the bottom holds about 5.8 mg/cm^2. Calculated on a mass/unit-volume basis, the respective values are 37 µg/ℓ (BNL) and 32 µg/ℓ (overlying water). In contrast to these vertically averaged values, the concentrations at the base of the BNL (20 m above the bottom) average 40 µg/ℓ compared to 28 µg/ℓ in the relatively "clear" water above the BNL. On the average, then, suspended sediment concentrations are about 1.5 times greater near the base of the BNL than in the directly overlying waters.

To estimate a residence time for these particles in the BNL, the rate at which they are being deposited must be known. Radiocarbon dating of the bottom sediments on Nitinat Fan gives an average net non-turbidite accumulation rate of about 6 mg/cm^2

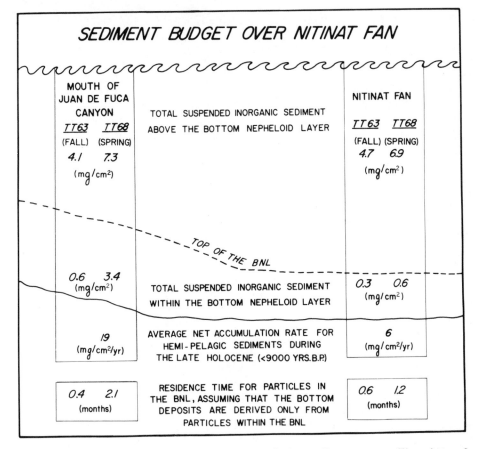

Fig. 11. Sediment budget for the Nitinat Fan area. The data for the canyon are from one station on each cruise, whereas the data for the fan represent 34 stations of cruise TT63 and 27 stations of cruise TT68.

throughout the late Holocene (0-9000 yrs. B. P.) [*Carson*, 1971]. Assuming that this accumulation is supplied only by particles from within the BNL, and that all of the particles within the BNL are eventually deposited on the portion of Nitinat Fan under consideration, then a mean residence time for these particles in the BNL is about 1 month.

Figure 11 shows the same calculations for a station at the mouth of Juan de Fuca Canyon. The late Holocene sedimentation rate used is from *Barnard* [1973]. The variance in residence time in this area is much larger, from 0.4 to 2.1 months, because of the appreciable differences in the mass of sediment within the BNL between the fall and spring cruises. The mean residence time, about 1.2 months, is not significantly different from that for Nitinat Fan proper.

Several important assumptions are implied in the preceding calculations: (1) that data from two cruises are representative of

the entire 9-month interval; (2) that radiocarbon estimates of
sedimentation rates based on 9000 yrs. of accumulation are valid
at present; and (3) that the particles undergo no resuspension
after deposition. Although these assumptions will certainly
undergo some modification as more data becomes available, a 1-month
residence time is probably realistic as a first approximation.
Eittreim and Ewing [1972], for example, calculate a maximum
residence time (assuming no resuspension) of 1 year for particles
in the nepheloid layer of the North American Basin, a layer
which is between 1 and 3 km thick.

The origin of the particles in the bottom nepheloid layers
found in most of the world's ocean basins is still conjectural.
Possibilities fall into two general categories: addition of particles from above the BNL supplied from turbidity currents, low-density
slope turbid layers, or pelagic sedimentation; and addition of
particles from below by various resuspension mechanisms (e.g.,
biological organisms, bottom currents). *Eittreim and Ewing* [1972]
favor injection of suspensates by gravity-controlled turbidity
currents as the principal source of particles for the 1-km thick
BNL found in the North American and Argentine Basins of the Western
Atlantic Ocean. The BNL over Nitinat Fan, however, is much
thinner, averaging about 125 m, and turbidity currents recognizable
in the sedimentary record have occurred with a periodicity of
only 500-600 yrs. during the late Holocene (0-9000 yrs. B. P.)
along the Washington and Oregon continental margins [*Griggs and
Kulm*, 1970; *Carson*, 1971; *Barnard*, 1973]. Furthermore, it seems
unreasonable to expect a large, infrequent turbidity current to
maintain a thin, relatively stable BNL. Continuously operative
processes, such as a steady influx of particles from the thick, low-
density bottom turbid layers recorded in the canyons on the slope
above Nitinat Fan [*Baker*, 1973], injection of sediment into the
water column by benthic organisms, or erosion by bottom currents,
appear to be more suitable sources of the Nitinat Fan BNL.

CONCLUSIONS

Sixty-five light-scattering profiles recorded during two
seasons over Nitinat Deep-Sea Fan have described an evidently
permanent and geographically continuous bottom nepheloid layer.
This layer is thinner than that found in many of the deep-ocean
basins, ranging between about 50 and 300 m. Scattering intensities
at the base of the BNL correlate well with the variations in
thickness, creating a relatively consistent picture in which the
thickest and most intense portions of the BNL occupy the topographic
lows (the northwestern flank and Cascadia Valley) leaving the
levees and bordering marginal ridge covered by only a relatively
thin and weak scattering layer.

Estimates of the amount of inorganic sediment which is
suspended within this layer were made on the basis of samples of
particulate matter collected *in situ* with the nephelometer readings.
The average value for the mass of inorganic sediment contained in a

1-cm^2 column of water within the BNL is about 0.5 mg. To sustain the net sedimentation rate recorded in Nitinat Fan over the last 9000 yrs., these particles must have a residence time of only about 1 month in the BNL.

REFERENCES

Ahlquist, N. C. and R. J. Charlson, Measurement of the vertical and horizontal profile of aerosol concentration in urban air with the integrating nephelometer, *Env. Sci. Tech*, *2*, 363-366, 1968.

Ahlquist, N. C. and R. J. Charlson, Measurement of the wavelength dependence of atmospheric extinction due to scatter, *Atm. Env.*, *3*, 551-564, 1969.

Baker, E. T., Nephelometry and Mineralogy of Suspended Particulate Matter in the Waters Over the Washington Continental Slope and Nitinat Deep-Sea Fan, Ph.D. Thesis, Univ. Washington, Seattle, 1973.

Barnard, W. D., Late Cenozic Sedimentation on the Washington Continental Slope, Ph.D. Thesis, Univ. Washington, Seattle, 1973.

Beutell, R. G. and A. W. Brewer, Instruments for the measurement of the visual range, *J. Sci. Inst.*, *26*, 357-359, 1949.

Carson, B., Stratigraphy and Depositional History of Quaternary Sediments in Northern Cascadia Basin and Juan de Fuca Abyssal Plain, northeast Pacific Ocean, Ph.D. Thesis, Univ. Washington, Seattle, 249, 1971.

Charlson, R. J., N. C. Ahlquist, H. Selvidge, and P. B. MacCready, Jr., Monitoring of atmospheric aerosol parameters with the integrating nephelometer, *J. Air Poll. Control Assoc.*, *19*, 937-942, 1969.

Eittreim, S. and M. Ewing, Suspended particulate matter in the deep waters of the North American Basin, in *Studies in Physical Oceanography*, vol. 2, edited by A. L. Gordon, pp. 123-167, Gordon and Breach, London, 1972.

Eittreim, S., A. L. Gordon, M. Ewing, E. M. Thorndike, and P. Bruckhausen, The nepheloid layer and observed bottom currents in the Indian-Pacific Antarctic Sea, in *Studies in Physical Oceanography*, vol. 2, edited by A. L. Gordon, pp. 19-35, Gordon and Breach, London, 1972.

Ewing, M., and S. D. Connary, Nepheloid layer in the North Pacific, in *Geological Investigations of the North Pacific*, Geol. Soc. Am. Mem. *126*, p. 41-82, 1970.

Ewing, M., S. L. Eittreim, J. I. Ewing, and X. LePichon, Sediment transport and distribution in the Argentine Basin, 3. Nepheloid layer and processes of sedimentation, in *Physics and Chemistry of the Earth*, vol. 8, pp. 49-77, Pergamon Press, New York, 1971.

Griggs, G. B., and L. D. Kulm, Sedimentation in Cascadia Deep Sea Channel, *Geol. Soc. Am. Bull.*, *81*, 1361-1364, 1970.

Hunkins, K., E. M. Thorndike, and G. Mathieu, Nepheloid layers and bottom currents in the Arctic Ocean, *Jour. Geophys. Res.*, *74*,

p. 6995-7008, 1969.

Ingraham, Jr., W. J., The Geostrophic Circulation and Distribution of Water Properties off the Coasts of Vancouver Island and Washington, Spring and Fall, 1963, *Fishery Bull.*, *66*, p. 223-250, 1967.

Jerlov, N. G., *Optical Oceanography*, Elsevier, Amsterdam, 194 p., 1968.

Sternberg, R. W., E. T. Baker, D. A. McManus, S. Smith, and D. R. Morrison, An integrating nephelometer for measuring suspended sediment concentrations in the deep sea, *Deep Sea Res.*, in press, 1974.

Thorndike, E. M. and M. Ewing, Photographic nephelometers for the deep sea, in *Deep Sea Photography*, edited by J. B. Hersey, Johns Hopkins University Press, Baltimore, Md., pp. 113-116, 1967.

Physical, Chemical, and Optical Measures of Suspended-Particle Concentrations: Their Intercomparison and Application to the West African Shelf

KENDALL L. CARDER, PETER R. BETZER, AND
DONALD W. EGGIMANN

University of South Florida

ABSTRACT

A property of oceanic particulate matter referred to as "apparent density" was calculated by dividing the weight of suspended particulate matter (SPM) by the volume of particles. This parameter is equal to "mass density" for particles, such as minerals, containing little water. Apparent density calculations were made for a series of samples collected on R/V Trident cruise 112 to the continental shelves of Sierra Leone and Liberia. These values ranged from 0.104 to 1.79 for samples with particulate organic carbon fractions (POC/SPM) ranging from 0.486 to 0.037.

Cross sections of salinity, light scattering β(45), suspended particulate matter (SPM), and β(45)/total surface area for this region of the west African shelf showed a northwestward-flowing bottom current laden with inorganic sediment having a high apparent density and s southeastward-flowing, organic-rich (low apparent density) surface current. Of the measures of particle concentration applied to these waters, SPM and β(45) showed greatest correlation (r = .960), suggesting that apparent density is highly correlated with the particle index of refraction. Total particulate volume and total particulate surface area data were not nearly as well correlated with either β(45) or SPM; optical/physical theories are proposed to explain this phenomenon.

INTRODUCTION

Optical, gravimetric, and electronic-sizing measurements have all been used by oceanographers to study the concentration and transportation of suspended particles in sea water [*Jerlov*, 1968; *Lisitzin*, 1972; *Sheldon et al.*, 1972]. To our knowledge no one has combined all three with measurements of particulate organic carbon and particulate carbonate; however, as will be shown, this combination can be used to provide estimates of the variability of the mean index of refraction and the mean specific gravity of suspended particulate matter. These properties are of paramount importance in studies of light propagation in the sea and particle dynamics within the water column. In this paper, some standard measures of particle concentration are compared and results of an initial investigation of the natural variations in the character of suspended particulate matter in a nepheloid region of the west African shelf using the named physical and chemical measurements are reported.

BACKGROUND

The three techniques compared — light scattering, electronic particle-sizing, and gravimetrics — are used to measure slightly different physical properties of suspended particles. Light scattering has, classically, been used for rapid assessment of relative concentrations of particles in waters too clear for accurate gravimetric or light attenuation measurements [*Jerlov*, 1968]. The measurement of oceanic particle-size distributions and volume concentrations has been greatly facilitated by employing automatic electronic sizing devices, such as the Coulter counter [*Sheldon and Parsons*, 1967]. The advent of large-volume, plastic water samplers (typically 30 to 1000 ℓ) and membrane filters has revolutionized gravimetric studies of suspended particulate matter in deep ocean regions where particle mass concentrations are normally < 20 mg/kg [*Spencer et al.*, 1970].

Light scattering by small spheres was shown by *Mie* [1908] to depend upon the size, index of refraction, and number of particles (for single-particle scattering) for incident light of given wave length and radiant intensity. *Hodkinson* [1963] showed that shape was not of major importance for randomly oriented, polydisperse distributions of particles. *Deirmendjian* [1963] showed that the volume scattering function β(45) (radiant intensity of light scattered at 45° relative to the irradiance incident upon a scattering volume) for a polydisperse distribution of non-absorbing spheres could be expressed theoretically, as

$$\beta(45) \propto \sum_{i=1}^{N} K_i D_i^2 \ , \tag{1}$$

where K_i is the scattering efficiency factor (proportion of the pencil of light geometrically obstructed by the ith particle to that scattered), D_i is the diameter of the ith particle, and N is the number of particles in the scattering volume. The expression for K_i can be written as

$$K_i = 2 + 4/R^2 \ (1 - \cos R) - 4/R \sin R \qquad (2)$$

where $R = \dfrac{2\pi D}{\lambda} (m - 1)$, with D the spherical particle diameter, λ the incident wavelength, and m the index of refraction of the particle relative to that of the medium. Figure 1 is a graphic

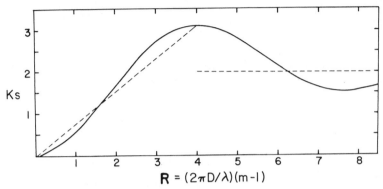

Fig. 1. Linear approximations to the scattering efficiency factor.

representation of this oscillatory function. The value of the relative index of refraction is pivotal for determining K_i since a typical value for organic particles might be 1.03 [*Carder et al.*, 1972; *Gordon and Brown*, 1972; *Zaneveld and Pak*, 1973] and for inorganic particles might be 1.17 [*Burt*, 1952; *Pavlov and Grechushnikov*, 1966]. The diameter D (4), corresponding to $R = 4$ (the first K_i peak), would vary from 1.50 μm ($m = 1.17$, e.g. inorganic particles) to 8.46 μm ($m = 1.03$, e.g. organic particles) for a wavelength of 0.4 μm in water. Consequently, K_i rapidly approaches a value of 2.0 as D increases for inorganic particle distributions while it increases more slowly with increases in D for organic particles. The dotted lines on Figure 1 represent a linear simplification of the efficiency factor curve resulting in the following equation:

$$K_i = \begin{cases} (2\pi D/\lambda) \ (m-1) \ (2.89/4), & R \leq 4 \text{ or } D \leq 2\lambda/(\pi(m-1)) \\[2mm] 2 & , \ R > 4 \text{ or } D > 2\lambda/(\pi(m-1)) \end{cases} \qquad (3)$$

where 2.89/4 is the slope of the line in the small-particle region of the curve, and K_i is constant for particles larger in optical size than $R=4$. Combination of (3) with (2) results in (4) and (5), providing simple dependences of light scattering on total particulate volume for optically small particles and total particulate surface area for optically large particles.

$$\beta(45) \doteq \begin{cases} k \cdot k_1 \sum_{i=1}^{N} D_i^3 = \dfrac{k(2.89)}{1.50} \dfrac{6(TV)}{\pi}, \text{ for } D_i \leq 1.50 \text{ } \mu m = D(4) \\[2em] k \cdot 2 \sum_{i=1}^{N} D_i^2 = 2k \dfrac{(TSA)}{\pi}, \text{ for } D_i > 1.50 \text{ } \mu m = D(4), \qquad (4) \\[2em] \text{assuming } m = 1.17 \text{ and } \lambda = 0.4 \text{ } \mu m \text{ in water;} \end{cases}$$

$$\beta(45) \doteq \begin{cases} k \cdot k_2 \sum_{i=1}^{N} D_i^3 = \dfrac{k(2.89)}{8.46} \dfrac{6(TV)}{\pi} \text{ , for } D_i \leq 8.46 \text{ } \mu m = D(4) \\[2em] k \cdot 2 \sum_{i=1}^{N} D_i^2 = 2k \dfrac{(TSA)}{\pi} \text{ , for } D_i > 8.46 \text{ } \mu m = D(4) \qquad (5) \\[2em] \text{assuming } m = 1.03 \text{ and } \lambda = 0.4 \text{ } \mu m \text{ in water,} \end{cases}$$

where k is a calibration factor, TV is the total particulate volume, and TSA is the total particulate surface area. In both cases, (assuming m is constant) the major variables are D_i and N. This indicates that light scattering by oceanic particulate matter is proportional to something less than the total particulate volume but more than the total particulate surface area, with the index of refraction of the particles determinging the slopes k_1 and k_2 as well as the critical diameter $D(4)$. For mixtures of organic and inorganic particles the problem is even more confusing. In general, distributions of refractive indices as well as particle sizes exist, and the use of "average" values for the distribution of indices has evolved [Zaneveld and Pak, 1973; Gordon and Brown, 1972] since particle-by-particle size and optical analysis is unfeasible at present. In deep nepheloid waters, where the inorganic fraction dominates, there is little variation in m and (4) approximates the scattering for refractive indices near 1.17.

Particle-size distributions are easily calculable with electronic devices for particle-size determination, such as the Coulter counter [Sheldon and Parsons, 1967; Carder et al., 1971; Brun-Cottan, 1971; Bader, 1970; Sheldon et al., 1972; Mulligan and Kingsbury, 1968; and Maloney et al., 1962]. This instrument measures cumulative particle-volume distributions from which spherical-equivalent cumulative distributions for the surface area of particles can be calculated.

A variety of shapes and conductivities characterize particles found in the oceans. Determining the sizes of such particles electronically may meet with some skepticism regarding accuracies in size measurement. Since the principle of measuring size using a Coulter counter involves drawing a particle through a constriction (orifice) in the electric field in a conducting medium (sea water), dielectric particles, by displacing a certain volume of the medium, produce increased potential pulses on passing through the

constriction. The size of these pulses is proportional to the individual volumes of the particles. The conductivity of a given
particle is unimportant as long as the conductivity is small compared with that of the medium. The high resistance of cell membranes
(e.g. *Nitella*) is thought to be due to the presence of a lipoidal
layer at the surface of the cell [*Giese*, 1963, p. 270] which, together with any cellulose, siliceous, or carbonaceous cell walls,
provide whole cells with high resistivities. Ruptured cells may result in decreased Coulter volumes due to the exposure of the cytoplasm to the medium. Since cytoplasm in such an instance might be
expected to be rather short-lived compared to the cell walls and
membranes, particle-volume measurement errors resulting from cells
transitory between whole cells and detritus are expected to be
small. Since natural marine geological particles are almost entirely dielectric, they also can be accurately measured. *Sheldon and
Parsons* [1967] defend the reliability of Coulter-counter data for
both marine organic and inorganic particles, provided precautions
regarding counting rates and cleanliness are followed. The accuracy
of their sand density measurements (2.6 versus 2.66 gm/ml) is an indication of the accuracy of the Coulter counter. In comparing the
Coulter counter to three other methods of size analysis for suspended sediments, *Swift et al.* [1972] found the Coulter counter to be the
most versatile method, with its major drawback to sedimentology
being that its size-determination data was based on a volume diameter rather than on an hydraulic diameter, which responds to shape
and roughness factors as well as to size. However, use of volume
diameter or spherical equivalent is not a drawback to hydrological
optics, since *Hodkinson* [1963] showed that a sphericity assumption
for Mie scattering was valid for polydisperse distributions of non-
spherical particles. Thus, use of spherical-equivalent surface
areas obtained using the Coulter counter is valid for comparisons
with scattering measurements. Spherical equivalents are not necessarily good assumptions for particle dynamics (e.g. settling velocities).

The dry mass of suspended particles, obtained by weighing, approximates the wet mass of most inorganic particles, but can differ
significantly from the wet mass of organic particles. Since about
90% of the composition of many phytoplankters is water, their dry
mass is quite small compared to their wet mass. Particle-volume
measurement and light scattering measure wet-particle characteristics so that their comparison with gravimetric determinations is
tenuous for these organic particles; however, little difficulty is
anticipated in making such comparisons for particle suspensions dominated by inorganic material (e. g. near-bottom nepheloid layers).

The suspended particulate matter *SPM* can be related to the total
particulate volume *TV* by the following expression:

$$SPM = \frac{\pi}{6} \sum_{i=1}^{N} (\rho_a)_i D_i^3 = \frac{(\sum_{i=1}^{N} (\rho_a)_i D_i^3)}{\sum_{i=1}^{N} D_i^3} (TV) = \overline{\rho}_a (TV) \quad (6)$$

where D_i is the ith particle diameter (spherical equivalent), N is the number of particles/mℓ, and $(\rho_a)_i$ is the "apparent" density of the ith particle, expressed as follows:

$$(\rho_a)_i = \frac{(m_d)_i}{(V_d + V_{H_2O})_i} \quad , \tag{7}$$

where m_d is the dry mass, V_{H_2O} is the volume of intracellular water before desiccation, and V_d is the volume of m_d. $\bar{\rho}_a$ is the mean "apparent" density of the suspended particles. The Coulter counter measures a wet volume $(V_d + V_{H_2O})$ and gravity techniques measure m_d. In contrast, the mass density ρ of a particle is expressed as

$$\rho = \frac{m_d + m_{H_2O}}{V_d + V_{H_2O}} \quad , \tag{8}$$

where m_{H_2O} represents the mass of intracellular water. For inorganic material $\rho = \bar{\rho}_a$ since V_{H_2O} and m_{H_2O} are, in most cases, small (6-14% H_2O by mass according to $Degens$ [1965])while ρ may be as large as $10\ \bar{\rho}_a$ for organic material.

For inorganic material one might expect m to increase as ρ increases. Figure 2 is a scattergram of the median values of the range of the relative refractive index and density listed by the *Handbook of Chemistry and Physics* [1966] for some common marine minerals. It shows that ρ and m are quite well correlated (0.89) but that some scattering of data does occur. The variabilities of m and ρ are small compared to that of $\bar{\rho}_a$ which might be expected to range from 0.1 (i.e., more than 90% of certain phytoplankters is water) to 2.93 (density of aragonite) making it the widest-ranging parameter discussed in this paper. For this reason, high correlations between SPM and TV are not expected unless the suspended particles studied are mostly inorganic material.

To identify some of the sources of variability among the various measures of particle concentration ($\beta(45)$, SPM, TV, TSA), it is important to measure the organic and the mineralogic fractions. Measurements of particulate organic carbon (POC) have been used [*Menzel and Vaccaro*, 1964] as a measure of the particulate organic content of seawater, while particulate carbonate (PC) can be used primarily to measure the calcareous fraction. The remaining fraction, ($SPM - POC - PC$) / SPM, consists of refractory substances, usually clay minerals -- montmorillonite, chlorite, illite, and kaolinite. Since PC and refractory components have indices of refraction and specific gravities that are quite high relative to those of phytoplankton and organic detritus [*Eppley et al.*, 1967; *Carder et al.*, 1972], inverse relationships between the organic fraction POC/SPM and both the apparent density ($\bar{\rho}_a$) and the relative index of refraction m are expected.

Measurements of the relative index of refraction of a multicomponent, polydisperse suspension of particles such as found in seawater have not been made. Estimates of the weighted mean value of the distribution of refractive indices have been made by $Zaneveld$

and Pak [1973] and *Gordon and Brown* [1972]; *Carder et al.* [1972] have estimated the relative index of refraction ($m = 1.03$) for a phytoplankter culture of *Isochrysis galbana*. As an indicator of

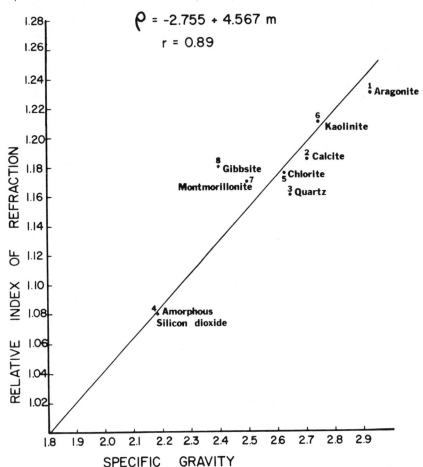

Fig. 2.   Relationship of the relative index of refraction to specific gravity for some inorganic particles.

changes in the relative index of refraction, the authors chose to apply the light-scattering vector ratio technique described by *Carder and Schlemmer* [1973], which is an extension of the light-scattering vector technique of *Pak et al.* [1970]. This ratio is formed by dividing $\beta(45)$ by *TSA*. It is essentially the light scattered per unit of surface area (spherical equivalent) of particles and is primarily affected by changes in m. *Carder and Schlemmer* [1973] found variations greater than fourfold in the Gulf of Mexico from upwelling regions (low-ratio values) to detrital areas (high-ratio values). The light-scattering vector ratio is presented in this paper to indicate variations in relative index of refraction.

EXPERIMENTAL PROGRAM

Measurements of light-scattering, particle-size distributions, suspended particulate matter, particulate organic carbon, particulate carbonate, and mineralogy were made on samples of suspended particulate matter of the west African shelf off Sierra Leone and Liberia during R/V *Trident* cruise TR-112. With the Guinea Current flowing southeastward near the shelf and with the possibility of an undercurrent flowing northwestward (suggested from *Hulbert and Thompson's* [1973] theory of upwelling) active erosion and sediment transport were anticipated as well as near-shore biological particle production. The suspended sediments carried along this shelf may reach the deep ocean with some regularity since the shelf is narrow (average width = 25 km) and is bounded on its seaward edge by numerous submarine canyons [*Egloff*, 1972].

Four hydrographic stations were made on each of fourteen transects of the shelf, extending over a distance of 480 km from near Shebro Island, Sierra Leone, to Cape Palmas in southeast Liberia. Since time available for sampling and hydro-bottle samples permitted *POC* measurements to be made at only seven stations, data from only these stations (Figure 3) are presented and discussed; these data represent the most complete ensemble of properties measured on the cruise. Suspended particles as indicators of circulation and sediment transport along the entire shelf-study region will be considered in subsequent papers.

Filtration time and the number of 30-ℓ bottles (eight) dictated the vertical resolution of the *SPM*, *POC*, and *PC* samples. At least two 30-ℓ samples were taken per station with as many as four at the deeper stations. The 30-ℓ samples were interspersed with 5-ℓ samples to provide temperature/depth data as well as additional light-scattering measurements used to interpolate particle concentrations. Salinity samples were obtained from all bottles. Vertical spacing was determined using the temperature structure provided by expendable bathythermographs (XBT's). At each deep (> 50 m) station, samples were gathered from surface waters, from the top of the thermocline, and from 4 m above the bottom.

Light scattering and particle sizes were measured *in vitro* aboard ship immediately after the hydro-bottle samples were obtained. Light-scattering measurements were obtained from all of the hydro bottles using a Brice-Phoenix photometer (Model 2000) [*Spilhaus*, 1965; *Pak*, 1970]; particle sizes between 2.3 and 20 µm were ascertained on samples from the 30-ℓ bottles using a Coulter counter Model B [described by *Carder and Schlemmer*, 1973; *Carder*, 1970; and *Sheldon and Parsons*, 1967] equipped with a 100-µm diameter aperture. *Pak* [1970] and *Carder* [1970] list absolute measurement errors for the Brice-Phoenix photometer and the Coulter counter, respectively, at less than 6% each.

Filtration was provided by a gravity-driven technique similar to that of *Betzer and Pilson* [1970], and was started as soon as the

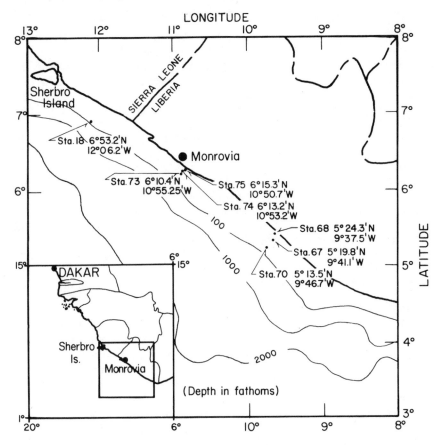

Fig. 3. Station locations for R/V *Trident* cruise TR-112.

bottles were aboard ship.

The 30-ℓ samples were filtered through 47-mm diameter Nuclepore 0.4 μ pore-size filters, Gelman glass fiber (Type A) filters,
and Millipore HA 0.45 μ pore-size filters. They were used,
respectively, for determination of suspended particulate matter
and particulate carbonate, particulate organic carbon, and clay
mineralogy. Normally about 4 ℓ of water was passed through the
Nuclepore and glass fiber filters while 12 ℓ was run through the
Millipore filters. Filtration was accomplished in a closed system:
the water was fed through a silicone rubber tube to an all-plastic,
in-line filter head (Millipore Filter Corp.). All filters were
handled with teflon tweezers (separate pairs were used to handle
each type of filter). After filtration, all filters were placed in
plastic Falcon tubes until processing could be carried out in a
shore-based "clean" laboratory. The Nuclepore filters were rinsed
twice in a clean bench (Baker Model 3220) with approximately 4 mℓ
doubly deionized water for each rinse. After the rinse water had
dripped through the filters, each was placed in an acid-rinsed

plastic tube, transferred to its own polycarbonate bottle containing silica gel, dried for two days and reweighed on a five-place Mettler balance (Model H20). Reweighing was accomplished in an air-conditioned room with a filtered air supply, in order to reduce possibility of contamination. The balance contained an *alpha* source to reduce static electric effects on filter weights [*Manheim et al.*, 1970]. The weight of suspended particles on each Nuclepore filter was calculated by substracting the tare weight (each had been weighed before the cruise on the same balance) from its desiccated weight and dividing by the volume of water which had been filtered.

Determinations of particulate organic carbon were made using an infra-red analyzer (Beckman Model 215A) according to the technique of *Menzel and Vaccaro* [1964]. The filters were oxidized in a Coleman Carbon-Hydrogen Analyzer (Model 33) at 690°C to prevent release of CO_2 from aragonite or calcite particles [*Fournier*, 1968]. No CO_2 was detected when samples of either aragonite or calcite were placed in the Carbon-Hydrogen Analyzer and run through the *POC* procedure at 690°C.

Determinations of particulate carbonate were made using a modification of the acetic acid dissolution technique of *Chester and Hughes* [1967]. In the present study 4 mℓ of 25% (v/v) acetic acid were added to each Nuclepore membrane supported in a plastic funnel having a Teflon stopcock at its base) and left for 2 hours. Following the acid treatment each filter pad was rinsed twice with 4 mℓ doubly deionized water for each rinse. The acid and rinse solutions were drained into 1-oz polyethylene bottles, transferred to 25-mℓ Class A Pyrex volumetric flasks and diluted to volume with doubly deionized water. The carbonate digestion procedure was checked using filtered samples of oceanic particulate matter from the Gulf of Mexico. The Nuclepore membranes on which the samples were collected were examined by microscope before and after treatment with acetic acid. The numerous coccoliths present in these samples were not detectable after exposure to acetic acid for 2 hours.

The solutions were analyzed for calcium and magnesium concentrations by atomic absorption spectrophotometry using a Perkin-Elmer unit (Model 403) equipped with a laminar flow burner and an air-acetylene flame. Samples were interspersed with calcium and magnesium standards and run in quadruplicate. In order to minimize viscosity effects on sample aspiration rates, the calcium and magnesium standards were prepared using the same acetic acid concentration that was used on the samples. Blank corrections to the calcium and magnesium determinations were made by leaching a Nuclepore filter with 25% (v/v) acetic acid for 2 hours and then analyzing the solution for its calcium and magnesium concentrations. The blank corrections were extremely small, averaging 0.6 µg and 0.05 µg for calcium and magnesium, respectively. The low calcium and magnesium levels in the acetic acid resulted in average signal-noise ratios of 12:1 and 9:1 for calcium and magnesium, respectively.

Calculations of *PC* concentrations for each sample were made by multiplying the concentration of the particulate calcium by 2.5.

This procedure assumes that the particulate carbonate is a pure $CaCO_3$; the high Ca/Mg ratios (10:1) from the samples studied indicated that this assumption is not unreasonable.

RESULTS AND DISCUSSION

In Table 1 are given properties of particles for the seven stations of cruise TR-112 at which particulate organic material was measured. Generally two or three different linear segments were found to fit the cumulative distribution data for size (spherical-equivalent diameter) plotted on log-log paper in a manner similar to that of *Bader* [1970], *Brun-Cottan* [1971], and *Carder and Schlemmer* [1973]. The "First Slope" column of Table 1 lists the log-log slope of the first, or small-particle, linear increment of the hyperbolic distributions fit to the particle data. This slope can be used to extrapolate into the small-particle region for estimates of the relative volumes and surface areas of particles smaller than those counted on cruise TR-112. These estimates can be used in calculating the "truncation errors" of statistics calculated from truncated particle distributions. The first slopes generally fall between 2.30 and 6.5 μm in diameter; *Carder and Schlemmer* [1973] suggested that small slopes were generally associated with upwelling regions while large slopes were generally associated with less productive regions. It should be noted that all particle-size parameters measured are given for particles larger than 2.3 μm in diameter.

Linear correlations between pairs of parameters from Table 1 are listed in Table 2. Samples excluded in the correlations are noted. The sample most often excluded in the comparisons was 73:25, because it appeared to be from a phytoplankton bloom region. This sample had high values of *TV*, *TSA*, and *POC/SPM* and low values of $\beta(45)$ and *SPM*. Its exclusion had only a small effect on the $\beta(45)$ versus *SPM* correlation, but a greater effect on $\beta(45)$ versus *TSA* and *TV* correlations, implying that $\beta(45)$ is a good indicator of *SPM*, particularly in regions dominated by inorganic particulate matter. Measurements made by the authors in the top 300 m of the equatorial Atlantic, where primarily biogenic particles were found (*POC/SPM* values as high as 0.80), resulted in the much lower correlation value of 0.81. Measures of total particle size (*TSA* or *TV*), on the other hand, can be highly distorted as measures of $\beta(45)$ or *SPM* if only one sample dominated by organic material, such as phytoplankton blooms, is included in the analysis.

Since the inclusion of the organic sample 73:25 had no significant effect on the correlation between $\beta(45)$ and *SPM*, m and ρ must be positively correlated. Using $\beta(45)/TSA$ and *SPM/TV* as indicators of m and ρ, respectively, results in a correlation coefficient of 0.741. Considering that *TSA* and *TV* are calculated from truncated distributions, this correlation is good.

The two primary subdivisions of suspended particles that can be determined chemically are the organic and inorganic materials, with the inorganic material having an appreciably higher

TABLE 1.   Particle Data from R/V *Trident* Cruise TR-112.

S	D (m)	SPM (μg/ℓ)	β(45) ($10^{-2}m^{-1}ster^{-1}$)	TV ($10^3\frac{μm^3}{mℓ}$)	TSA ($10^3\frac{μm^2}{mℓ}$)	First Slope (-)	β45/TSA ($10^{-2}ster^{-1}$)	POC (μg/ℓ)	POC/SPM	SPM/TV ($g/mℓ^3$)	N ($mℓ^{-1}$)
75:	10	46.4	1.16	191	168	3.23	6.905	12.5	.269	.243	3703
	36	474.0	4.33	968	710	2.31	6.10	87.7	.185	.490	11349
74:	22	58.2	1.32	357	288	2.69	4.60	19.0	.327	.163	5203
	37	133.8	1.62	415	330	2.73	4.95	23.4	.175	.322	5983
73:	25	82.6	.967	795	526	2.25	1.85	40.1	.486	.104	8280
	55	58.1	1.08	210	188	3.00	5.75	20.6	.355	.277	3645
	80	101.9	1.35	182	172	2.83	7.85	27.3	.268	.560	3458
70:	45	114.6	1.72	174	200	3.31	8.65	--	--	.659	4810
	65	51.9	.912	62	70	3.00	12.90	--	--	.837	1608
	90	109.1	1.21	96	96	3.35	12.50	--	--	1.14	2167
67:	12	27.6	.880	226	192	2.99	4.55	12.4	.449	.122	3614
	27	40.9	1.20	182	198	3.05	6.05	12.6	.308	.225	4289
	42	48.5	1.06	136	156	3.12	6.80	10.3	.212	.357	3810
	63	664.1	4.31	372	466	3.05	9.25	47.3	.071	1.79	11209
68:	12	--	1.13	184	178	2.79	6.30	--	--	--	3592
	42	523.6	3.24	354	378	2.44	8.55	38.4	.073	1.48	7654
18:	17	43.3	1.10	280	186	2.64	5.90	20.2	.467	.155	3433
	34	54.5	.923	132	118	3.08	7.85	--	--	.413	2359
	54	62.1	.735	88	90	2.68	8.20	--	--	.706	1860
	74	150.6	1.28	86	110	3.49	11.70	5.6	.037	1.75	2825

S = Station Number
D = Sample Depth

TABLE 2. Linear Correlations Among Some Particle Properties.

Parameters	r	N	X
SPM vs. TV	0.840	16	73:25, 67:63, 68:42
	0.601	18	73:25
	0.464	19	——
β_{45} vs. TSA	0.905	19	73:25
	0.751	20	——
β_{45} vs. SPM	0.961	18	73:25
	0.960	19	——
β_{45} vs. TV	0.768	19	73:25
	0.577	20	——
β_{45}/TSA vs. SPM/TV	0.741	19	——
β_{45}/TSA vs. POC/SPM	-0.806	14	——
SPM/TV vs. POC/SPM	-0.600	14	——

r = the linear correlation coefficient
N = the number of points used
X = samples excluded in the correlations

density and index of refraction than the organic material. Comparisons of $\beta(45)/TSA$ and SPM/TV values to POC/SPM values in Table 2 result in correlation coefficients of -0.806 and -0.600, respectively. Certainly the subdivision of suspended particles into their organic and inorganic fractions is a step toward delimiting m and ρ, and the relatively high negative correlation of $\beta(45)/TSA$ with POC/SPM provides the researcher with a field tool for rapid identification of organically or inorganically dominated suspended particulate samples. This information justifies the assumption of *Carder and Schlemmer* [1973], based upon work by *Pak et al.* [1970], that $\beta(45)/TSA$ can be used to identify oceanic upwelling areas.

The correlation of SPM to TV is dependent upon which samples are excluded. Correlation is dominated by large values, so the early linear sense is established by sample 75:36. Relative to this sample, the other large values are from samples of highly inorganic material (samples 67:63 and 68:42) and one sample of organic material (sample 73:25). The correlation coefficients given in Table 2 reflect the decrease in correlation resulting from their inclusion. As noted, poor correlation results between SPM and TV if ρ is quite variable. On the African shelf, ρ_a ranged from 0.104 to 1.79, resulting in a rather low correlation coefficient (0.464) between SPM and TV.

The larger correlation coefficient for total surface area versus $\beta(45)$ as compared to total volume versus $\beta(45)$ suggests that $\beta(45)$ is better represented by a surface-area function for the samples from the African shelf. Equations (3) and (4) indicate, then, that the particles measured should be predominantly inorganic material.

Examination of Table 1 corroborates this expectation since the data
is dominated by samples containing greater than 70% inorganic material by weight. The most glaring exception to this statement is
found in sample 73:25 (48.6% POC) which also has a low optical vector ratio value ($\beta(45)/TSA$). This sample, occurring some 60 m
above the bottom and in the euphotic zone, is apparently comprised
predominantly of phytoplankton and detritus.

Figure 4b is a scattergram of $\beta(45)$ versus TSA containing, also,
the best fitting least-squares linear regression curve through these
points. Sample 73:25 is shown circled, but was omitted in the calculation of the least-squares line. The relatively high correlation coefficient 0.905 dropped to 0.751 when sample 73:25 was included, illustrating the effect the inclusion of a blooming phytoplankton sample has on the correlation. The term "phytoplankton"
is used rather than "organic" since there can be a drastic variation in m_{H_2O} depending upon whether a sample contains live phyto-

plankton or organic detritus.

Figure 4b is a scattergram of $\beta(45)$ versus SPM. A comparison
of equations (4) and (6) suggests that $\beta(45)$ and SPM should have
high correlation in the following cases:

(a)   When the particles are small organic material ($D < 8.46\mu m$)
        with constant refractive indices and apparent densities
        (variable N and size distributions), $\beta(45)$ is a volume-
        type function;

(b)   When the particles are large inorganic material ($D > 1.5$
        μm), and variations in the size distributions are small
        (i.e. $TSA \propto TV$; and

(c)   When m is proportional to ρ, and the variability of N is
        large compared with other property variations.

The high correlation (0.96) between $\beta(45)$ and SPM is, obviously, not
due to (a) above, but is, rather, probably due to a combination of

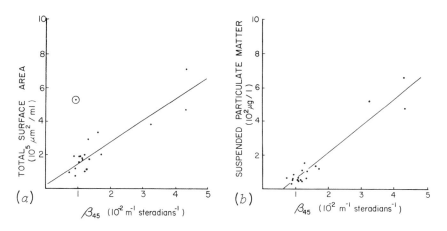

Fig. 4.   Relationship of $\beta 45$ to (a) total surface area with the
least-squares regression line (circled data point excluded), and to
(b) suspended particulate matter with the least-squares regression
line.

(b) and (c). Since there is a significant variation in the shapes of the particle-size distributions ("First Slopes" in Table 1) and the particles are, for the most part, inorganic material by weight, the effect of $\beta(45)$ being for the most part an area-type function for inorganic material and SPM being a volume type of function must be compensated by the wide variablity of N and the rather good correlation (.74) between $\bar{\rho}_a$ and $\beta(45)/TSA$.

A comparison of the light-scattering vector ratio $\beta(45)/TSA$ with the apparent density $\bar{\rho}_a$ is given in Figure 5. These indicators of the refractive index for particles and density, respectively, are reasonably well correlated ($r = 0.74$) considering that the former indicator is a wet-particle measure, while the latter is a combination of a dry SPM- and a wet TV-particle measure. Variations in the volume of the intracellular water within the organic fraction (e.g. living phytoplankton may contain much larger amounts of intracellular water than does organic detritus) can greatly distort $\bar{\rho}_a$ compared to ρ according to equations (6) and (7). The good

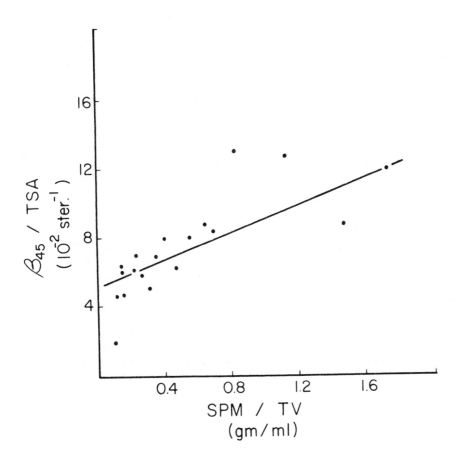

Fig. 5. Relationship of $\beta(45)/TSA$ versus SPM/TV or $\bar{\rho}_a$ with the least-squares regression line.

agreement between $\bar{\rho}_a$ and $\beta(45)/TSA$ helps to support (c) above.

The variability of the First Slope (Table 1) ranges from 2.25 to 3.79. While this does not seem to be an extraordinarily wide range, it does have significant effect upon the truncation fractions of TV and TSA. Using the hyperbolic distribution, the First Slope can be extrapolated back to 0.5 µm to estimate the fractions of TSA and TV truncated by counting only those particles of diameter > 2.3 µm. To extrapolate it much further would be unrealistic since the number of particles approaches infinity for diameters approaching 0. Using a log-log slope of 2.25 results in TSA and TV truncation fractions of 0.04 and 0.27, respectively; for a slope of 3.49, the fractions become 0.28 and 0.83, respectively. Such truncation results in increased point scattering in Figures $4a$ and 5 since distributional changes vary the proportion of light-scattering particle mass due to occurence of particles that are too small to be represented in TV and TSA. The only real solution to this truncation problem is to measure the smallest particles possible.

Of the three cases for which high correlation between SPM and $\beta(45)$ is expected, a combination of (b) and (c) seems most applicable to the African shelf samples. In the first place, the particles are not primarily organic material and have quite variable indicators of refractive index and apparent density. Secondly, although $\beta(45)/TSA$ and $\bar{\rho}_a$ were well correlated, the variability of the size distributions does not allow TV and TSA to be related by a simple constant. Thirdly, although the variability of N is rather high, if $\bar{\rho}_a$ and $\beta(45)/TSA$ were not so well related, the correlation between $\beta(45)$ and SPM would not have been as high as 0.96.

The greatest correlation between $\beta(45)$ and SPM is expected for deep nepheloid layers where the organic content is very low and where $\bar{\rho}_a = \rho$. In such cases, as shown in Figure 2, a high degree of correlation is also expected between m and ρ. The correlation may be somewhat decreased, because $\beta(45)$ will be more closely related to TSA than TV for deep nepheloid layers, but the high variability of N, the correlation of m to ρ, and the expected similarity in the shapes of size distributions due to a paucity of organic material, should result in very high correlations. On the other hand, significantly lower correlations are expected for regions having narrow ranges of N, wide variations in $\bar{\rho}_a$ and m, and significant ranges in the shape of the particle-size distributions. The top 300 m of the open-ocean water column might represent such a region where a phytoplankton-dominated assemblage of particles near the surface changes to a detritus-dominated assemblage of particles with depth.

Application of the above measures of particle concentration was made to particle transport in the waters of the west African shelf. Line 10 of Figures 6 and 7 was selected as typical of most transects made along the shelf in terms of both its hydro-dynamics and its suspended sediment characteristics. Figure $6a$ is a cross section along Line 10, depicting the salinity structure of the shelf. A current meter was placed at near-bottom between

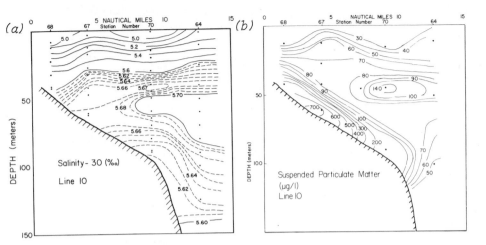

Fig.6.   Cross section of Line 10 for the Liberian shelf: (a) salinity and (b) suspended particulate matter.

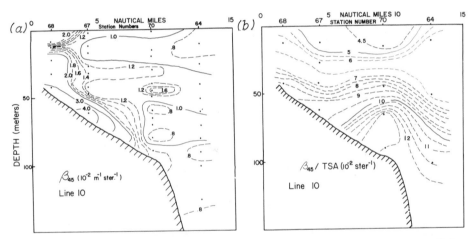

Fig. 7.   Cross section of Line 10 for the Liberian shelf: (a) light scattering and (b)light-scattering vector ratio.

stations 67 and 70 (Figure 6a). The kaolinite-rich bottom layer ($z > 50$ m) flowed predominantly northwestward while the montmorillonite-rich surface layer (Guinea Current) flowed southeastward [*McMaster et al.*, in preparation, 1973]. Judging from the halocline, the shear zone was probably between 40 and 50 m depth.

Figures 6b, 7a, and 7b are cross sections of *SPM*, ß(45) and ß(45)/*TSA*, respectively. They show a nepheloid layer, probably generated by scouring effects of bottom currents which at times exceed 20 cm/sec. ß(45) and *SPM* show similar turbidity distributions with more structure apparent in the former due to its higher

sampling density. The ratio $\beta(45)/TSA$ reveals the increasingly
inorganic character of the suspended particles as bottom is
approached. Low values of $\beta(45)/TSA$ in the surface layer imply
active phytoplankton productivity there similar to that suggested
by *Carder and Schlemmer* [1973]. The maximum-salinity layer is
quite clean except where it contacts the bottom and its relatively
high values of $\beta(45)/TSA$ suggest that detritus is probably its
predominant particle-type except near the bottom. The relative
particle-concentration maximum of sample 70:45 is probably
stability-induced since it occurs at the depth of the maximum-
density gradient in accordance with the findings of *Jerlov* [1959],
Carder et al. [1971], and others.

CONCLUSIONS

A property of oceanic particulate matter referred to here as
"apparent density" was calculated by dividing the weight of
suspended particulate matter by the volume of particles. This
parameter is equal to "mass density" for particles (e.g. minerals)
containing little water. Apparent density calculations were
made for a series of samples collected on R/V *Trident* cruise 112
to the continental shelf regions of Sierra Leone and Liberia. These
values ranged from 0.104 to 1.79 for samples with particulate
organic carbon fractions (POC/SPM) ranging from 0.486 to 0.037.
Of the measures of particle concentration applied to the
waters of the west African shelf off Liberia, SPM and $\beta(45)$
were the best correlated ($r = 0.960$), suggesting that apparent
density $\bar{\rho}_a$ is highly correlated with the index of refraction m.
Total particulate volume and total particulate surface area
were not nearly as well correlated with either SPM or $\beta(45)$.
The effect of varying values of the particle-mass density and index
of refraction served to decrease the above correlations.
Indicators of these, SPM/TV and $\beta(45)/TSA$, respectively, were
well correlated ($r = 0.74$) as were $\beta(45)/TSA$ and POC/SPM ($r = -0.80$),
while SPM/TV and POC/SPM were only moderately correlated ($r = 0.600$).
A simple linear light-scattering model was presented with
particle-size and scattering data to show the differences between
light scattering from organic and inorganic particles. The model
suggested that the light scattering from the suspended particles of
the Liberian shelf was dominated by light scattering from inorganic
material.   This implication was confirmed by measurements of the
contribution of organic and inorganic materials to the suspended
particulate matter.
The relatively high correlation (0.80) of $\beta(45)/TSA$ with $POC/$
SPM provides an important diagnostic field tool for identification
of organic particles. Regions of high productivity or bottom
scour can be identified by use of a light-scattering photometer and
a Coulter counter within an hour of recovery of hydro-bottle
samples. This technique should enable certain sampling strategy
decisions to be made relatively rapidly in the field.

Cross sections of salinity, light scattering, suspended particulate matter, and $\beta(45)/TSA$ indicated a northwestward flowing, sediment-laden bottom current and a southeastward flowing surface current containing a large organic particle fraction. Comparison of POC/SPM and SPM/TV data with the trends of the $\beta(45)/TSA$ cross section indicates the salient features of each is well represented by $\beta(45)/TSA$.

ACKNOWLEDGEMENTS

This research was supported by the Office of Naval Research through Contract N00014-72-0363-0001. The helpful suggestions of and discussions with Dr. Susan B. Betzer and Scott I. McClelland are appreciated. We were ably assisted during our sampling program by the officers and crew of the University of Rhode Island's R/V *Trident*.

REFERENCES

Bader, H., The hyperbolic distribution of particle sizes, *J. Geophys. Res.*, *75*, 2822, 1970.

Betzer, P. R. and M. E. Q. Pilson, Concentrations of particulate iron in Atlantic open-ocean water. *J. Mar. Res.*, *28*, 251, 1970.

Brun-Cottan, J. C., Etude de la granulométrie des particles marines measures effectuées avec un Compteur Coulter, *Cah. Oceanogr.*, *23* (2), 193, 1971.

Burt, W. V., Scattering of light in turbid water, Ph.D. thesis, University of California, Los Angeles, 1952.

Carder, K. L., Particles in the eastern equatorial Pacific Ocean: Their distribution and effect upon optical parameters, Ph.D. thesis, Oregon State University, Corvallis, 1970.

Carder, K. L., G. F. Beardsley, Jr., and H. Pak, Particle size distribution in the eastern equatorial Pacific, *J. Geophys. Res.*, *76*, 5070, 1971.

Carder, K. L., R. D. Tomlinson, and G. F. Beardsley, Jr., A technique for the estimation of indices of refraction of marine phytoplankters, *Limnol. Oceanogr.*, *17*, 833, 1972.

Carder, K. L. and F. C. Schlemmer II, Distribution of particles in the surface waters of the eastern Gulf of Mexico: An indicator of circulation, *J. Geophys. Res.*, *78*, 6286, 1973.

Chester, R. and M. J. Hughes, A chemical technique for the separation of ferro-manganese minerals, carbonate minerals and adsorbed trace elements from pelagic sediments, *Chem. Geol.*, *2*, 249, 1967.

Degens, E. T., *Geochemistry of Sediments, A Brief Survey*, Prentice-Hall, Englewood Cliffs, New Jersey, 342 pp., 1965.

Deirmendjan, D., Scattering and polarization properties of polydispersed suspensions with partial absorption, in *I.C.E.S. Electromagnetic Scatterings*, vol. 5, edited by M. Kerker, pp. 171-189, Pergamon, London, 1963.

Egloff, J., Morphology of ocean basin seaward of northwest Africa:

Canary Islands to Monrovia, Liberia, *Amer. Assoc. Petrol. Geol. Bull., 56*, 694, 1972.

Eppley, R. W., R. W. Holmes and J. D. H. Strickland, Sinking rates of marine phytoplankton measured with a fluorometer, *J. Exp. Mar. Biol. Ecol., 1*, 191, 1967.

Fournier, R. O., Observations of particulate organic carbon in the Mediterranean Sea and their relevance to the deep-living coccolithophorid *cycloccolithus fragilis, Limnol. Oceanogr., 13*, 693, 1968.

Giese, A. C., *Cell Physiology*, second ed., W. B. Saunders Press, Philadelphia, p. 270, 1963.

Gordon, H. R. and O. B. Brown, A theoretical model of light scattering by Sargasso Sea particulates, *Limnol. and Oceanogr., 17*, 826, 1972.

Handbook of Chemistry and Physics, 47th ed., Chemical Rubber Co., Cleveland, 1966.

Hodkinson, J. R., Light scattering and extinction by irregular particles larger than the wavelength, in *I.C.E.S. Electromagnetic Scattering*, vol. 5, edited by M. Kerker, pp. 87-100, Pergamon, London, 1963.

Hulbert, H. E. and J. D. Thompson, Coastal upwelling on a β-plane, *J. Phys. Oceanogr., 3*, 16, 1973.

Jerlov, N. G., Maxima in the vertical distribution of particles in the sea, *Deep Sea Res., 5*, 178, 1959.

Jerlov, N. G., *Optical Oceanography*, Elsevier, Amsterdam, 194 pp., 1968.

Lisitzin, A. P., Sedimentation in the World Ocean, *Soc. Econ. Paleontol. and Mineral., Spec. Pub. 17*, Tulsa, 218 pp., 1972.

McMaster, R. L., P. R. Betzer, K. L. Carder,   L. Miller, and D. W. Eggimann., Suspended particle mineralogy and water masses of the west African shelf adjacent to Sierra Leone and Liberia, in preparation, 1973.

Maloney, T. E., E. J. Donovan, Jr., and E. L. Robinson, Determination of numbers and sizes of algal cells with an electronic particle counter, *Phycologia, 2*(1), 2, 1962.

Manheim, F. T., R. H. Meade, and G. C. Bond, Suspended matter in surface waters of the Atlantic margin from Cape Cod to the Florida Keys, *Science, 167*, 371, 1970.

Menzel, D. W. and R. F. Vaccaro, The measurement of dissolved organic and particulate carbon in sea water. *Limnol. Oceanogr., 9*, 138, 1964.

Mie, G., Beiträge zur Optik trüber Medien, speziell kolloidalen Metallosüngen, *Ann. Physik, 25*, 377, 1908.

Mulligan, H. F. and J. M. Kingsbury, Application of an electronic particle counter in analyzing natural populations of phytoplankton, *Limnol. Oceanogr., 13*, 499, 1968.

Pak, H., The Columbia River as a source of marine light-scattering particles, Ph.D. thesis, Oregon State University, Corvallis, 1970.

Pak, H., G. F. Beardsley, Jr., G. R. Heath and H. Curl, Light scattering vectors of some marine particles, *Limnol. Oceanogr., 15*, 683, 1970.

Pavlov, V. M. and B. N. Grechushnikov, Some aspects of the theory of daylight polarization in the sea, *U. S. Dept. Comm., Joint Pub. Res. Ser., Rept. 36*(816), 25, 1966.

Sheldon, R. W. and T. R. Parsons, *A Practical Manual on the Use of the Coulter Counter in Marine Science*, Coulter Electronics, Toronto, 66 pp., 1967.

Sheldon, R. W., A. Prakash, and W. H. Sutcliffe, Jr., The size distribution of particles in the ocean, *Limnol. Oceanogr., 17*, 327, 1972.

Spencer, D. W., D. E. Robertson, K. K. Turekian and T. R. Folsom, Trace element calibrations and profiles at the Geosecs test station in the northeast Pacific Ocean, *J. Geophys. Res., 75*, 7688, 1970.

Spilhaus, A. F., Jr., Observations of light scattering in sea water, Ph.D. thesis, Massachusetts Institute of Technology, Cambridge, Mass., 1965.

Swift, D. J. P., J. R. Schubel, and R. W. Sheldon, Size analysis of fine-grained suspended sediments: A review, *J. Sed. Petrology, 42*(1), 122, 1972.

Zaneveld, J. R. V. and H. Pak, Method for the determination of the index of refraction of particles suspended in the ocean, *J. Opt. Soc. Amer., 63*(3), 321, 1973.

The Distribution of Particulate Matter in a Northwest African Coastal Upwelling Area

GUNNAR KULLENBERG

University of Copenhagen

ABSTRACT

The particle distribution in the upper 200 m of the upwelling area oceanward of the northwest African coast has been investigated, using an in situ integrating light-scattering meter, with the two-fold purpose of relating to the general conditions in the area regarding circulation, topography, biology, and chemistry, and investigating the possibility of defining and tracing water masses in such an area by their particle content. The investigation demonstrates that the particle content of water can serve as an indicator of physical and biological processes. Particle content very distinctly characterizes water-type. The great advance in using in situ light-scattering instruments combined with simultaneous temperature recording to study the oceanic particle distribution is apparent.

OBSERVATIONS

The observations on which the present study is based were carried out during the period 2-9 March 1972 in the area 19°45'N 16°58'W to 19°15'N 17°50'W, approximately 40 nautical miles northwest of Cape Timiris. The work was done on board the research vessel *Meteor*. The topography of the area proved to be complicated, with several small canyons cutting the shelf line (Figure 1). In all, 40 stations, distributed in sections of 15-20 nautical miles in length and running approximately at right angles to the general direction of the coast line (Figure 1), were occupied. Observations were repeated at several sections so that possible variability could be studied.

195

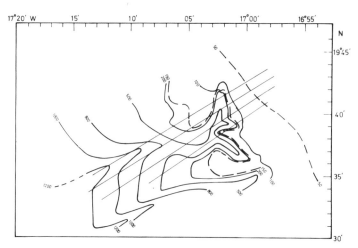

Fig. 1. Topography of the southern portion of the study area.
The thin lines indicate positions of sections.

     The observational program at each station included temperature/
salinity soundings using a Bathysonde (German STD, standard procedure), chemical observations using inorganic and organic com-
pounds (various techniques), and optical measurements. The temp-
erature/salinity and chemical observations were carried out by
scientists from Kiel and Hamburg. The optical program was comprised of simultaneous light-scattering and temperature profiles at
0-200 m depths using an *in situ* integrating light-scattering meter
equipped with a rapid response thermistor and a depth gauge. The
instrument consists of a diffused light source with a photomultiplier tube used as detector [*Jerlov*, 1961; *Jerlov*, 1968, p. 19 ff].
The instrument measures total scatterance in the red (655 nm)
part of the spectrum (Schott and Genossen filter RG 610) to obtain
a measure of particle content. In this end of the spectrum, scatterance due to the water itself is negligible relative to particle
scatterance.
     The scattering meter was calibrated to give total scattering
coefficient $b(km^{-1})$ (using the method described by *Jerlov*, 1961)
and total amount of suspended matter in mg/ℓ. The latter calibration was carried out by adding to a given volume of water known
amounts of suspended material obtained by resuspension of undryed
Baltic sediments. Since water and suspended matter from the area
in which the scattering measurements were made could not be used
for the calibration, this data must be regarded with some caution.
     Weather conditions during the measurement period were favora-
ble, with northerly winds varying between 4 and 16 m/sec, implying
presence of conditions for coastal upwelling. According to the
mechanism of upwelling dynamics, oceanic water penetrating toward
the coast at some intermediate depth, upwelling in a narrow, more
or less frontlike zone, and relatively cold nutrient-rich surface
water drifting slowly toward the ocean outside the upwelling zone

should then be expected [*Sverdrup*, 1937; *Smith*, 1968; *Hidaka*, 1972].

RESULTS AND DISCUSSION

The density (σ_t) and particle distributions for two typical sections are shown in Figures 2 and 3. Contours drawn in the figures are not labeled since they conform to both density and particle distribution. The arrows in the two figures indicate the author's interpretation of the water movements. Three essentially different water masses characterized by certain ranges of temperature, salinity, and particle content can be defined (Table 1) as clear oceanic water, old upwelled water, and shallow water. The mass of clear oceanic water situated below the surface layer outside the upwelling zone can be characterized as having low temperature and being poor in suspended particles. The mass of old upwelled water occupying the surface layer outside the upwelling zone can be characterized as having relatively low temperature and high

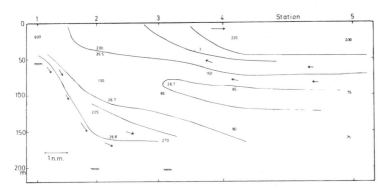

Fig. 2. Section from shallow to deep water on 6 and 7 March, *Meteor* stations 149 (shallow) to 153 (deep). Numbers indicate density (σ_t) (e. g. 26.5) and scattering in relative units (e.g. 70).

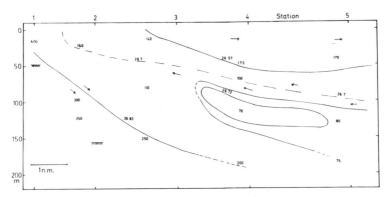

Fig. 3. Section from deep to shallow water on 8 March, *Meteor* stations 157 (deep) to 160 (shallow). Numbers indicate density (σ_t) (e.g. 26.5) and scattering in relative units (e.g. 70).

TABLE 1.   Characteristics of the Water Masses off the Northwest
Coast of Africa.

Water Masses	S°/00	T°C	σ_t	Particle content (mg/ℓ)
Clear oceanic water	35.4 - 35.5	12.9-13.4	~26.7	0.06-0.11
Old upwelled water	35.75- 35.65	14.6-15.6	~26.4	0.2 -0.4
Shallow water	36.0 - 36.8	15.5-16.8	26.6-27.	1.2 -5.0

particle content. The shallow water mass, or bank water, inside the
upwelling zone with depths less than 50 to 60 m can be characterized
as turbid, warm and saline.

The circulation in the two sections shown in Figures 2 and 3
can reasonably be interpreted as follows. Clear, relatively cold
oceanic water drifts slowly toward the coast beneath the surface
layer in the outer parts of the sections. The upwelling seems to
be concentrated in a rather narrow zone centered around station 2.
A small part of the upwelling water is probably mixed into the bank
water, but the dominating part of it returns to the sea as a slow,
outward surface drift. This surface layer occupies the top 50 m
outside station 3. As the nutrient-rich upwelled water returns to
the sea in the surface layer, its temperature and particle content
are increased. The particle increase is more rapid than the temperature increase and can be explained by enhanced primary production
due to the high nutrient content.

The presence of the shallow water mass, or bank water, compli-
cates the interpretation of the circulation. It is not unreasonable
to assume that the upwelling creates a barrier which prevents the
bank water from spreading outward toward the sea. During the dry
season, the water is trapped on the bank and, as a consequence, the
temperature and salinity as well as particle content are increased
during the dry season.

Beneath the clear ocean water, at stations 2,3 and 4, an inversion with warm, saline water of high particle content is observed.
This water can only originate from the bank water. Since a small
part of the upwelling water is probably mixed into the bank water,
a return flow of bank water toward the sea must occur. This return
flow can only occur beneath the upwelling; the observed inversions
are, therefore, interpreted as outflowing bank water. While the
water is flowing outward, it is mixed, to a certain degree, with
water from above and, farther out, possibly from below.

At least to the 200-m depth contour, the outflowing bank water
forms a bottom current. The density and particle content of the
bottom layer in the shallow water mass are very high (Table 1), and
the outgoing current is explained, for the most part, as a gravitational effect. The present slope in the depth range 50-250 m is
2° to 3° (Figure 1). The order of magnitude of the velocities in
the bottom current due to gravitational effects can be estimated
by

$$\tau_b = c_d \cdot \rho U^2 = g \cdot \Delta\rho \cdot H \cdot \sin\phi$$

where τ_b = bottom stress, c_d = drag coefficient, ρ = density, U = velocity, g = acceleration of gravity, $\Delta\rho$ = density-excess relative to the ambient water, H = hydraulic mean thickness of the current, and ϕ = slope.

The following values can be obtained from the observations: $\sin\phi = 4.5 \cdot 10^{-2}$, $H \approx 10$ m, and $5 \cdot 10^{-5} \le \Delta\rho \le 3 \cdot 10^{-4}$. The range of the drag coefficient can be estimated from near-bottom current observations in coastal waters [*Bowden*, 1962] and from the experiments on turbidity currents by *Kuenen* [1951], cited by *Kullenberg* [1954]. A reasonable range is $3 \cdot 10^{-3} \le c_d \le 3 \cdot 10^{-2}$. With these values the order of magnitude of the current velocity is 20-25 cm sec^{-1}.

The maximum velocity can be estimated by the relationship

$$U_m = C \cdot (H \cdot g \cdot \frac{\Delta\rho}{\rho} \sin\phi)^{\frac{1}{2}}$$

where the numerical factor C is about 10 [*Tesaker*, 1969]. This yields the value 37 cm $\cdot$ sec^{-1}. These velocities are not unreasonable in the light of continuity requirements. They are sufficiently large to prevent the suspended matter from settling and to cause a certain amount of bottom erosion. Occasional bursts of outflowing water can cause considerable erosion.

It should be pointed out that the high salinity of the shallow water mass is the main source of the high density of the water.

The overall description of the sections is subject to considerable variability in details. This is evident when profiles obtained at different times at approximately the same positions are compared. Figures 4, 5 and 6 show profiles from positions at the inner, central and outer parts of the sections. During the period of observation, the clear oceanic water was encountered at depths between 50 and 120 m. The tongue of clear water rising toward the surface in the upwelling zone was found between 30 and 75 m depth.

The particle content of the different water masses is also subject to variations (Table 1). It is noted that the overall range of the particle concentraton is much larger than the temperature and salinity ranges. The present estimates of the total

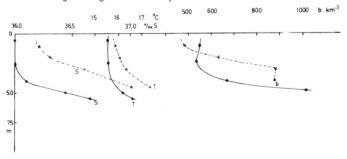

Fig. 4.   Examples of profiles in shallow water.
          o = *Meteor* station 149, 6 March 2220 hours
          x = *Meteor* station 160, 8 March 1010 hours

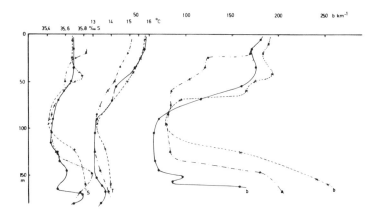

Fig. 5.   Profiles from the central parts of the sections.
          x = *Meteor* station 151, 7 March 0230 hours
          ● = *Meteor* station 157, 8 March 0310 hours
          Δ = *Meteor* station 163, 8 March 2310 hours

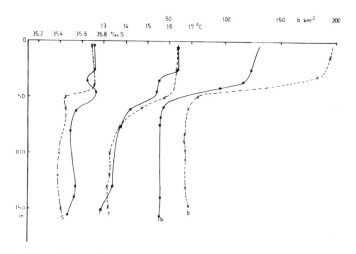

Fig. 6.   Profiles from the outer parts of the sections.
          o = *Meteor* station 134, 4 March 0840 hours
          x = *Meteor* station 145, 6 March 0215 hours

amount of suspended matter for the offshore water conform rather
well with recent investigations, such as those by *Jacobs and Ewing*
[1969]. Further, the amount of suspended matter found in the shallow water mass is in reasonable agreement with other estimates
[*Jerlov*, 1968, p. 26].

     The present observations can be compared with those of *Pak
et al.* [1970] in the upwelling area off the Oregon coast. The distribution of the particulate matter found there is similar to that
presented here. It is interesting to note that a similar inversion
layer of high turbidity was also observed off the Oregon coast.

The generation mechanism for this layer differs, however, from the mechanism presented here.

CONCLUSIONS

This investigation demonstrates the great advantages of using *in situ* light-scattering measurements, preferably combined with simultaneous recording of temperature and salinity, to study the oceanic particle content and distribution. It is also shown that the particle distribution in an upwelling area can serve as an indicator of physical and other processes and that the particle content is a very distinct characteristic of a water mass. Of special interest is the inversion layer of high particle content found beneath the upwelling water. The high particle load of the bottom water on the slope indicates the presence of an eroding current.

ACKNOWLEDGEMENTS

The author wishes to express his thanks for the invitation to participate in the cruise and to acknowledge the excellent cooperation with the staff on board the research vessel *Meteor*. Thanks are also due to Mr. N. Berg Olsen who assisted during the measurements. Support from the Danish Natural Science Research Council is acknowledged.

REFERENCES

Bowden, K., Measurements of turbulence near the sea bed in a tidal current, *J. Geophys. Res.*, *67*, 3181-3186, 1962.

Hidaka, K., Physical oceanography of upwelling, *Geoforum*, *11*, 9-21, 1972.

Jacobs, M. B., and M. Ewing, Suspended particulate matter: Concentration in the major oceans, *Science*, *163*, 380-383, 1969.

Jerlov, N. G., Optical measurements in the eastern North Atlantic, *Medd. Oceanog. Inst. Goteborg, Ser. B, 8*, 40 pp., 1961.

Jerlov, N. G., *Optical Oceanography*, Elsevier, Amsterdam, 194 pp., 1968.

Kuenen, Ph. H., Properties of turbidity currents of high density, *Soc. Econ. Paleont. Mineral.*, *Spec. Publ. 2, 1*, 1951.

Kullenberg, B., Remarks on the Grand Banks turbidity current, *Deep Sea Res.*, *1*, 203-210, 1954.

Pak, H., G. F. Beardsley, Jr., and R. L. Smith, An optical and hydrographic study of a temperature inversion off Oregon during upwelling, *J. Geophys. Res.*, *75*, 629-636, 1970.

Smith, R. L., Upwelling, in *Oceanogr. Mar. Biol. Ann. Rev.*, vol. 6, edited by H. Barnes, pp. 11-46, 1968.

Sverdrup, H. U., On the process of upwelling, *J. Mar. Res.*, *1*, 155-164, 1937.

Tesaker, E., Uniform turbidity current experiments, *Inst. Hydr. Constructions*, Techn. Univ. Norway, 40 pp., 1969.

The Suspended Material of the Amazon Shelf and Tropical Atlantic Ocean

RONALD J. GIBBS

University of Delaware

ABSTRACT

Filtered samples and optical data from six cruises in the Atlantic Ocean off the Amazon River between 1963 and 1971 are used to determine transport of surface suspended material outward into the ocean. During high-river discharge the turbid-water line (≥ 2.0 mg/ℓ) extends 100 km seaward from the river mouth and northwestward along the coast for about 2000 km in a zone averaging 60 km wide. During low-river discharge, a similar pattern extends seaward only 80 km from the river mouth and northwestward for about 2000 km. This turbid zone migrates between these extreme limits at intermediate stages of river discharge. The surface concentration of suspended material exhibits a steady seaward decrease and the zone along the outer continental shelf shows lobes of turbid water being engulfed by the northwestward flowing Guiana current.

INTRODUCTION

The overall purpose of the present study is to determine the mechanisms of transport of, and the *en route* changes that have taken place to, the sediments discharged into the tropical Atlantic Ocean from a point source, the Amazon River. The portion of the investigation reported here is concerned mainly with the distribution of the concentration of the surface suspended material on the Amazon shelf and the adjacent tropical Atlantic Ocean area. A detailed study of the vertical distribution is in progress in the area extending 800 km offshore, that is, outward to the depths of the Guiana Basin.

STUDY AREA AND PROCEDURES

The study area in the tropical Atlantic Ocean is shown in Figure 1. The main source of the detrital suspended material of the area is

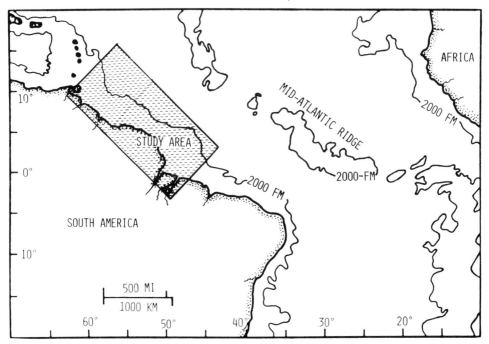

Fig. 1.   Location of study area.

the Amazon River, with almost negligible amounts being supplied by the smaller rivers (the Maroni, Essequibo, and Courantyne) to the northwest.   The surface circulation is dominated by the north-westward flowing Guiana current on the outer shelf in the deep-water portions of the study area. Longshore circulation likewise flows northwestward, and is induced primarily from the wave pattern strik-ing the coast.

These waves also produce turbulent shear on the bottom of the shallow portion of the shelf which is, actually, an extensive part of the shelf area. The large tide range (up to 10 m at the mouth of the Amazon) adds to the available energy of the environment with the continual movement of water on and off the shelf area. The overall combination of the Guiana current, longshore transport, tidal currents and waves produces an environment of high energy for the transportation of sediments.

Locations of surface suspended material samples are shown in Figure 2 for high-river discharge and in Figure 3 for low-river discharge. Field work was accomplished between 1963 and 1971 on several cruises into the area on the R/V *Chain* of Woods Hole Oceano-graphic Institute, the R/V *Almirante Saldanha* and the *Bertioga* of

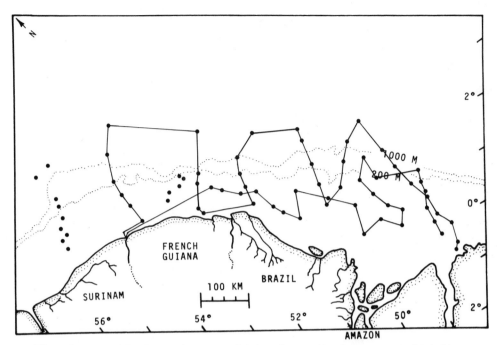

Fig. 2. Sample locations at high-river discharge. Solid line and station on it show continuous optical experiment of 1971 *Almirante Saldanha* cruise.

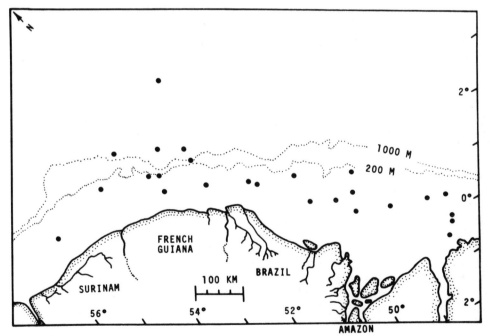

Fig. 3. Sample locations at low-river discharge.

the Brazilian Navy, the R/V *Oregon I* of the U.S. Bureau of Commercial Fisheries, and a tugboat of the Companhia de Indústria e
Comércio de Minas Geraís (ICOMI). The track of the continuous optical transmission and scattering study of the surface area is shown
on Figure 2 as a solid line.

On the more recent cruises, water samples were taken using a
special pump which pumped seawater directly into the pressure-
filtering chambers at a rate of 40 ℓ/min. In this way the samples
were at no time exposed to air. Samples were taken between 1 and
5 m below the surface in order to avoid floating contaminants. On
the earlier cruises, samples were taken just below the surface with
a specially cleaned bucket or with a bag sampler. The bag sampler
is collapsed before sampling then extends to engulf the sample when
triggered.

The samples were pressure-filtered through pre-weighed 1.45μ
pore-diameter filters. The volume of water filtered varied from
20-ℓ samples in high-concentration areas to 250-ℓ samples in low-
concentration areas. All traces of sea salt were removed by washing
and dialysis before the filtered samples were weighed. During
weighing of all filters and filtered samples, humidity was maintained
at 40% and blank filters were processed along with the filtered samples in order to determine filter blank effect.

The optical study was carried out on water samples which were
pumped aboard continuously into a special system having a 50-cm
light path for transmission observation. The simultaneous optical
scattering was observed in the same cell using a 90° scattering angle
and a photomultiplier. Calibration curves were constructed by filtering sets of samples after optical readings were obtained on splits
of the same sample.

RESULTS AND DISCUSSION

The surface distribution of the concentration of suspended
solids in the ocean area affected by the Amazon River during high-
river discharge is shown in Figure 4. The high-river discharge of
the Amazon River has a mean concentration of 123 mg/ℓ and discharge
of 268 x 10^3 m³/sec. The area of high concentration extends in a
continuous band along the shore northwestward to the Gulf of Paria.
The width of the zone extending out as far as the 5 mg/ℓ line varies
from nearly 100 km to about 10 km in such other areas as that off
French Guiana. This high-concentration zone was sampled from ships
and was also observed from aircraft flights over the area. Seaward
of the high-concentration zone, the boundaries take on a lobate
aspect especially prevalent between the 0.2 to 1 mg/ℓ boundaries.
These lobes are discussed by *Gibbs* (1970), who studied the salinity
and temperature distribution of the area. It appears that the Guiana
current is shearing off the large lobes and engulfing them as it
flows swiftly past the turbid brackish water.

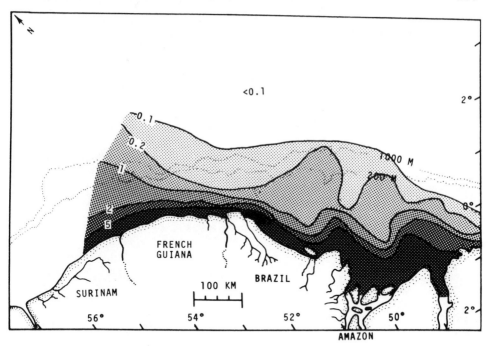

Fig. 4. Concentrations of surface suspended material (mg/ℓ) at high-river discharge.

During low-river discharge, the Amazon River has a mean concentration of 22 ppm and discharges 95×10^3 m³/sec. This is approximately 18% of the concentration of high-river water discharge and 6% of the solid discharge. The overall pattern of distribution for low-river discharge is similar to that for high-river discharge (Figure 5) except that the turbid water does not extend as far seaward. The turbid zone migrates between these two extreme limits at intermediate stages of river discharge.

The data from the point samples on the stations shown on Figures 4 and 5 provide an array of data points which are then contoured. The optical study of the surface waters, on the other hand, is a continuous data record. This continuous line of data shows that much more structural variation exists in the concentration of suspended material than the point-sampling method can possibly determine. The significant additional information gained from the optical data (Figure 6) indicates lineations along the coast of French Guiana and Surinam and many lobes on the seaward edge of the turbid zone off the Amazon River. It was also discovered that instead of a steady gradient from river mouth to open ocean, there exist numerous high and low ridges and valleys within the gradient. The majority of these lineations are parallel to the direction of flow and are even visible from satellite photographs.

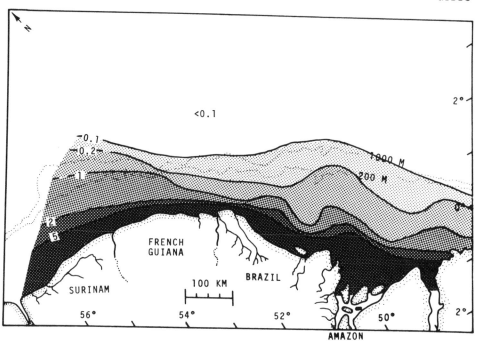

Fig. 5.   Concentrations of surface suspended material (mg/ℓ) at low-river discharge.

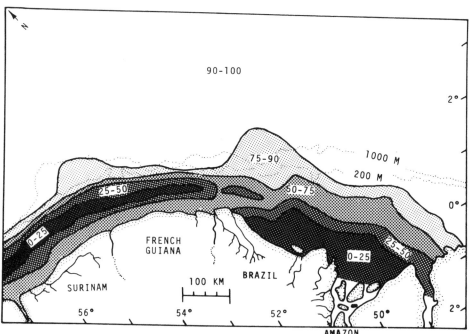

Fig. 6.   Percentage of optical transmission of surface waters during high-river discharge.

The vertical distribution at all these stations is being compiled along with the current structure for this area and exhibits a complex relationship. While the variation of the surface concentration associated with a tidal cycle may change as much as twofold, the concentration of deeper layers shows changes as great as 200 fold.

In a recent study of the bottom sediments on the Amazon shelf, *Gibbs* [1973] found a zone of high concentration of $<2\mu$ bottom sediments corresponding to the area of high surface concentration of suspended material found in the present study. This association suggests that the obvious source for the mud zone of bottom sediments is the high concentration of suspended material coming from the Amazon River.

The observed patterns of distribution can represent two different phenomena: (1) advection, or transport by currents, and (2) diffusion outward from a point source. Superimposed on these two distinct phenomena, we must consider that the suspended material in a given water mass is not a conservative property since, as it moves along, concentration, size distribution, and composition may change. To a certain extent, the finest material could be considered the most conservative property because of its slow settling velocity in relation to the prevailing turbulence. However, even the conservative nature of this fine material is not demonstrated completely since this material could flocculate to form particles of much higher settling velocity or the material could be ingested by creatures and defecated as fecal material of, again, much higher settling velocity. Not only can material be removed from the water mass but material can also be added from other sources, such as biological growth, to further complicate the understanding. Therefore, suspended material could change with time for a diffusion model as well as for an advection model. In the case of the advection model, as a parcel of water is moved, likewise all the suspended sediment would be moved with it to its new location and would, *en route*, be sedimenting material as a function of settling velocity. The diffusion model is more complicated since not all the material is transported. In this model, the material actually diffuses at different rates for the different size fractions of suspended material, as well as sedimenting from the water mass at different rates. To determine which of these two models is valid, we must have both data on the suspended material and information on the movement of the water. This knowledge of the water currents suggests the advective term, with the remainder suggesting the diffusive term. Presently, the structure of the water current in this area is being compiled. It is a complex structure. At this stage of the investigation, the advective term appears to dominate the distribution, as represented by the lobes extending seaward, which could not be due to diffusion, and by the movement along the shore in only the northwestward direction. If diffusion in the horizontal sense is present, it could only be accounting for the gradients extending seaward from these obvious advective-caused structures. However, even these gradients could be caused by sedimenting material from an advecting water mass, in which case diffusion would be insignificant in this study as a mechanism of transport.

The noteworthy conclusions of the investigation of the suspended material discharged into the tropical Atlantic Ocean from the Amazon River, a point source, is (1) that the material is dominantly transported northwestward along the shore for almost 2000 km, (2) that the surface concentration of the suspended material exhibits a steady decrease seaward, and (3) that the zone along the outer continental shelf shows lobes of turbid water being engulfed by the northwestward flowing Guiana current. This tracing of the suspended material from the Amazon River northwestward demonstrates, for the first time, that the Amazon River definitely does supply the material for the accretion of sediments to the shore. This is in agreement with the geomorphological theory of accretion to the coast line proposed by *Reyne* [1961]. The lobes and lenses of higher concentrations of suspended material seaward of the continental shelf are compatible with the lobes and lenses of low salinity observed in this area by *Gibbs* [1970]. These lobes and lenses, along with the general transport of the suspended material, are the mechanisms for the surface movement of the suspended sediment oceanward.

ACKNOWLEDGEMENTS

I wish to thank the officers and crews of the ships from which the field work was accomplished. Appreciation is extended to my research assistants, David Ellis and Bruce Weber, for aid in the analytical work. The United States Office of Naval Research provided the funds for this work under ONR Contract N00014-67-A-0356-0011.

REFERENCES

Gibbs, R. J., Bottom sediments of the Amazon shelf and tropical Atlantic Ocean, *Mar. Geol., 14,* M35-M45, 1973.
Gibbs, R. J., Circulation in the Amazon River estuary and adjacent Atlantic Ocean, *J. of Mar. Res., 28,* 113-123, 1970.
Reyne, A., On the contribution of the Amazon River to accretion of the coast of the Guianas, *Geol. Mijnbouw., 40,* 219-226, 1961.

IV

OFFSHORE STUDIES

Turbidity Distribution in the Deep Waters of the Western Atlantic Trough

Stephen L. Eittreim and Maurice Ewing

Lamont-Doherty Geological Observatory
University of Texas

ABSTRACT

The "background" turbidity (clearest water of the water column) has been mapped in the western Atlantic using standardized nephelometer measurements taken on R/V Conrad cruises 15 and 16. This background has then been utilized as a reference turbidity to map features of the bottom-water turbidity using older unstandardized relative nephelometer profiles. Variations in background turbidity of a factor of 3 apparently reflect the pattern of biological productivity in the overlying surface waters: high background in an equatorial belt and in latitudes greater than the subtropical convergences, but low background in the temperate latitudes. The resulting distributions of near-bottom turbidity values show a maximum in the southwestern Argentine Basin and a lesser maximum along the continental rise of the North American Basin. Bottom turbidity in the nepheloid layer decreases from both basins toward the equator and appears to be both a function of bottom current velocity and proximity to terrigenous sediment sources.

INTRODUCTION

It is well known that the western Atlantic Ocean is the site of intensive exchange of polar bottom waters. The high velocities of these thermohaline boundary currents and the interaction of this moving water with the bottom sediments result in a turbid bottom boundary layer (nepheloid layer) in most places along the Western Atlantic Trough [*Ewing and Thorndike*, 1965; *Eittreim et al.*, 1969; *Ewing et al.*, 1971; *Eittreim and Ewing*, 1972]. This layer can be thought of as analogous to a turbid river flowing over a mud bottom

213

with particles constantly being deposited and/or eroded at the interface. Along the path, local sources may add suspended particles. A large number of nephelometer profiles, taken on various cruises of the R/V *Vema* and R/V *Conrad* using the Ewing-Thorndike photographic nephelometer, have recorded this nepheloid layer. Most of these measurements could be used as relative vertical profiles only, due to a lack of standardization of the nephelometer readings. During the past two years, standardized nephelometers have been used on *Conrad* cruises 15 and 16, *Vema* cruises 29 and 30, and the R/V *Knorr* leg of GEOSECS from Dakar to New York. Portions of *Conrad* cruises 15 and 16 traversed the length of the western Atlantic from Bermuda to the Scotia Sea giving good coverage with the standardized measurements. In this paper, these measurements from *Conrad* cruises 15 and 16 are used to correlate the previous data using a reference turbidity level to give a mappable data coverage of 286 stations in this most interesting region.

This paper is a preliminary report on an ongoing study whose desired end result is to identify the sources and transport paths of sediments to the deep sea and along the basin deeps. The conversion of the turbidity measurements reported here to amounts of sediments in suspension can be accomplished by empirical comparisons of nephelometer measurements with suspended sediment concentrations collected at the time of measurement.

INSTRUMENTATION

An up-to-date description of the nephelometer used, its units of measurement and its precision of measurement are given in *Thorndike* [in preparation 1973], and more general descriptions are given in *Eittreim et al.* [1969] and *Eittreim and Ewing* [1972]. The instrument consists of a camera with continuously transported 35-mm film which records exposures from forward-angle scattering in the water (approximately 8 through 24 degrees) and from an attenuated, direct beam. Since the transmittance factor (absorption) of the opal-glass/gray-filter attenuator is fixed and is measured before and after cruises, the ratio of film exposure E resulting from scattered light to that of the directed attenuated light E_D gives a measure of the light scattered from particles in the water. This assumes that changes in direct beam attenuation by the water itself are negligible over the one-half meter path length.

According to *Matlack* [1972], the maximum attenuation at a wavelength of 478 nm in the bottom nepheloid layer of the Hatteras Abyssal Plain is .12m^{-1} which, for a path length of one-half meter, would attenuate the beam by about 5%. The absorption of the direct beam by the opal-glass/gray-filter attenuator attenuates the beam by a factor of 10^3. The transmittance of the attenuator was chosen to give a light level approximating that of scattering from deep water of the Sargasso Sea. This deep water is known to be extremely low in scatterance and thus establishes, for convenience, a zero-level turbidity roughly relating to a known water mass. By using the

ratio E/E_D, instrumental changes such as changes in the film's
transport speed and its sensitivity or film development are compensated for and the ratio is a function of the degree of scattering
produced by suspended particles. Scattering from the water itself
is negligible compared to the particle scattering at these forward
angles [*Kullenberg*, 1968].

A bourdon-tube pressure recorder records depths by deflection
of a light spot on the film. The film's transport speed, the slit
width and the lowering rate combine to limit the depth-resolution of
the system. Under normal operating conditions, a given point on the
film actually represents the average of scattering over a depth
range of approximately 20 m. Film exposures are derived from density measurements on the film combined with sensitometer data, which
are a series of controlled exposure steps recorded on the film just
prior to descent. The precision of the measurement is dependent on
both the precision of the density measurement and that of the conversion to log exposure through the relationship of the log E versus
film density obtained with the sensitometer. The estimate of this
precision is ±.05 log E.

Unfortunately, only the more recent data have been reducible to
these standard units since the transmittance of the direct-beam attenuator was not always held constant, nor were shipboard modifications of the equipment adequately recorded on cruises before 1972.
Thus, only the more recent data for which the attenuator has been
fixed or altered in a known fashion are used to map absolute turbidity changes.

The units of turbidity can be related to approximate concentrations of suspended matter through the empirical relationship shown
in Figure 10 of *Biscaye and Eittreim* [this volume]. Concentrations
sampled in the North American Basin are plotted versus the scattering value ($S=E/E_D$) recorded for the depth at which the water sample
was taken. On the basis of this relationship, scattering values,
log E/E_D, of 0.7, 1.0 and 1.6 represent concentrations of 8, 20,
and 120 mg/ℓ, respectively. Caution is urged, however, in broadly
applying this relationship to other regions of the ocean where the
particulate composition may differ sufficiently to produce a different relationship.

MEASUREMENTS

In order to utilize the large amount of data taken prior to the
standardized measurements, a reference turbidity level in the water
column is used. The geographic variations of this reference turbidity level must be known over the region to be mapped and then unstandardized data must be referred to that level. It is assumed
that time variations (seasonal) in the reference turbidity level are
sufficiently smaller than the geographic variations to be ignored.
Evidence of seasonal variations and their effect on the method used
here are discussed below.

Most nephelometer profiles in the deep basins of the world show rather steady decreases in turbidity from the surface downward, with generally higher gradients and, often, minor maximums associated with thermocline regions. An increase is then encountered, starting a few hundred meters above bottom. This is the bottom nepheloid layer [*Ewing and Thorndike*, 1965]. Thus, a broad minimum in turbidity usually occurs in the intermediate-to-deep waters [*Eittreim et al.*, 1969; *Eittreim and Ewing*, 1972; *Eittreim et al.*, 1972; and *Biscaye and Eittreim*, Figure 3, this volume]. This minimum, or clearest water, shall be referred to here as the "background" turbidity. This background turbidity has been used as a reference level in past work (see the references given above) for studies where local constancy was assumed. In these studies, only data from those stations reasonably distant from the edge of the continental shelf were used and, since the variations in scattering in the bottom water were great, it was felt the assumption was sufficient to determine the major features of the bottom nepheloid layer. However, it is, admittedly, a tenuous assumption and with use of this background turbidity as a reference over long distances and crossing oceanographic boundaries, any variations of turbidity that might exist must be determined.

Of the 286 stations from which data are used, 175 profiles were made with the standardized nephelometers. For the most part, these were stations from *Conrad* cruises 15 and 16, with 3 stations from *Knorr* cruise 30 (Figure 1). Only data from those stations deeper than 3000 m and west of the Mid-Atlantic Ridge were used. Unfortunately, to date no new measurements have been made north of the Gulf Stream in the northern North American Basin.

CLEAREST-WATER OR BACKGROUND TURBIDITY

The variations in background turbidity shown in Figure 2 range from about 0 to 0.4 log E/E_D, or a change in turbidity by a factor of nearly 3. The high values occur in regions of high biologic productivity south of the subtropical convergence (approximately 40°S) and in the equatorial region from about 5°S to 10°N. The low values occur in the biologically barren temperate latitudes centered at about 20°N and S. This strongly suggests that the background turbidity of the water column is simply a function of the rate of fallout of biogenic scatterers from the surface waters [*Be and Tolderlund*, 1971].

The latitudinal dependence of the background turbidity is illustrated for both cruises 15 and 16 of the R/V *Conrad* in Figure 3. In addition to the latitudinal factor (productivity variations), proximity to the continent is also important and accounts for the principal differences in the data on the two cruises. For example, the stations of cruise 16 at 10° to 20°S were taken far from the continent (Figure 3) whereas those of cruise 15 at the same latitude were taken close to the continental margin (Figure 1). Similarly, the group of station points of cruise 16 at 30°N were taken along

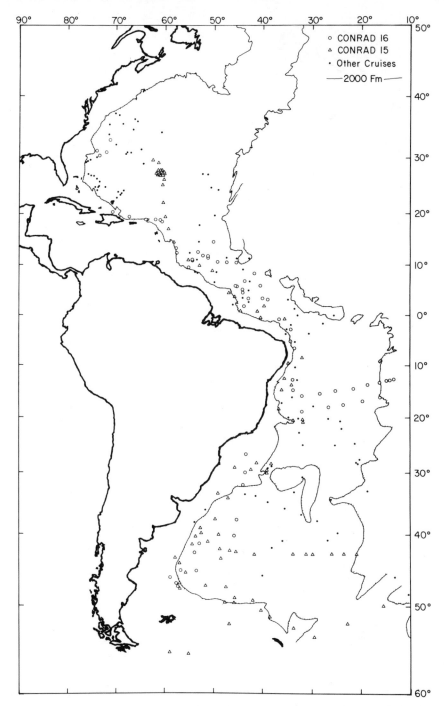

Fig. 1. Location of stations Δ and O = stations at which data was taken using the standardized nephelometer. The three stations centered at 30°N, 73°W from the R/V *Conrad* cruise 16, 20 profiles from which average values were taken.

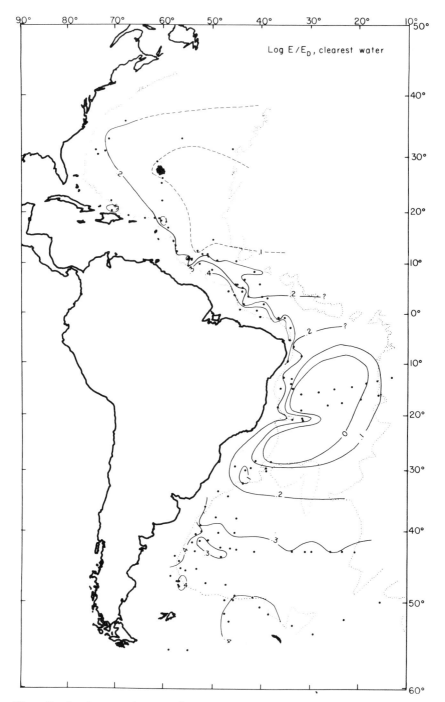

Fig. 2. Background, or clearest-water, turbidity values. Units
are $\log_{10}$ ratio of film exposure on the scattering trace to that on
the direct attenuated trace. Dotted line represents the 2000-fm
isobath.

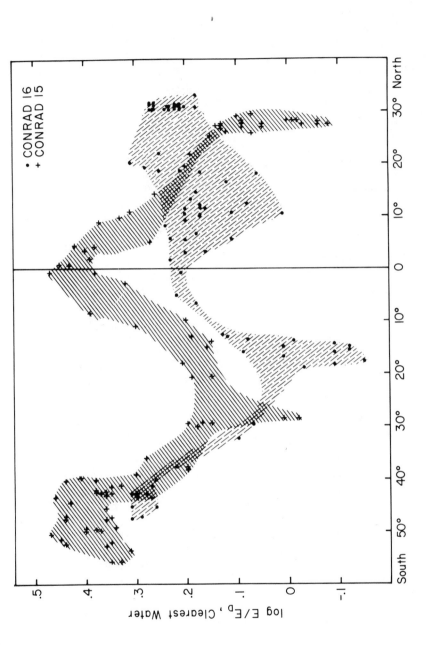

Fig. 3. Minimum turbidity value (background turbidity) versus latitude, R/V *Conrad* cruises 15 and 16.

the continental margin of the North American Basin whereas the data
from cruise 15 at similar latitudes were taken far from the continent at the southeastern edge of the Bermuda Rise. Despite the
effect of proximity to continent, which can be seen in the contour
interpretation of Figure 2, the strong latitudinal dependence still
produces a similarity in trends of background turbidity versus latitude for the two cruises, illustrated in Figure 3, with lowest
values recorded in the regions of lowest productivity.

At the equator, the difference in background-turbidity values
for the two cruises may be due to seasonal changes, although the
stations of *Conrad* cruise 15 at which the higher turbidities were
recorded were, again, on the average, closer to the continental margin than the stations of cruise 16. Seasonal factors are suggested
by the fact that the higher values, from cruise 15, were recorded in
June and the lower values, from cruise 16, were recorded in October.
Studies of seasonal phytoplankton production in the equatorial
Pacific show maximum production in April and a minimum production
in October [*Owen and Zeitzschel*, 1970]. Thus a difference might be
expected in the rate of fallout of organic detritus between the
months of June and October in the equatorial Atlantic, assuming a
similar seasonal dependence exists here.

The origin of the equatorial high in background turbidity is
complicated by the possible contribution of suspended sediment from
the Amazon to the deep-water column north of the equator [*Gibbs*, this
volume]. Owing to the Guiana Current, sweeping northwestward along
the coast, the bulk of the Amazon material is carried in that direction. Hence, this should not be considered as a source south of
about 2°N. This analogy suggests that the data points located south
of 2°N with values greater than 0.3 cannot be explained by this
source and, therefore, must be due, we believe, to the equatorial
high-productivity zone. Recent data from R/V *Conrad* cruise 3, not
included in the figures here, confirm the existence of this equatorial high in background turbidity well south and east of the Amazon
River.

The variations in background turbidity observed in Figure 2 are
gradual and regular. Sharp gradients are infrequent and no convolu-
ted contouring was necessary to represent all the points properly.
From this map, one can judge the precision of an estimate of background turbidity for any given location. An estimate of this precision would vary depending on the location; in the areas of higher
gradients it might be as high as ±0.2 log E/E_D or, in other areas,
about ±0.1. If the difference in equatorial values recorded is indeed a seasonal effect as discussed above, such an effect would
produce errors of ±.2 in this region (Figure 3).

BOTTOM-WATER TURBIDITY

The values plotted on Figure 4 are derived from film exposures
which represent scattering at a height of about 40 m above the

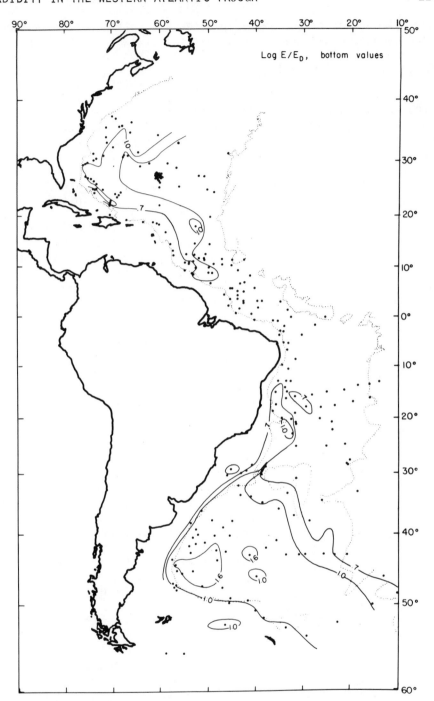

Fig. 4. Near-bottom turbidity values. 40% of the data stations utilize the background turbidity as a reference turbidity level which is assigned a value based on Figure 2. Dotted line represents the 2000-fm isobath.

bottom. About 40% of the values plotted are obtained from unstandardized nephelometer data taken on earlier cruises and utilize the
background turbidity as a reference level. Thus, a conservative
estimate of precision of the values on this map would be ±0.2, based
on the precision with which the background turbidity is known.

The variations in turbidity encountered from station to station
for the bottom-water values (Figure 4) are far greater than the variations of the background turbidity (Figure 2). The bottom-water
turbidity decreases toward the equator from the western North
American Basin on the north and the Argentine Basin on the south.
The bottom water of the Argentine Basin has significantly higher
turbidity than that in the North American Basin. The highest turbidity values in the Argentine Basin are encountered in the southwestern portion where the Antarctic Bottom Water (AABW) passes the
southernmost canyons of the Argentine continental slope [*Lonardi and
Ewing*, 1971]. This suggests the possibility that new suspended material is being added through the canyons to the already turbid AABW
which flows into the basin from the south through a gap in the Falkland Fracture Zone at 49°S, 36°W [*Le Pichon et al.*, 1971]. The
pattern of Figure 4 indicates that, although the bottom waters of
the western North American Basin (North Atlantic Deep Water) and the
Argentine Basin (AABW) are relatively turbid, most of the particles
drop out of suspension within these basins and only minor amounts
are carried the full length of the Western Atlantic Trough by the
AABW. The possibility that the particles are diffused upward,
rather than falling out of suspension, can be ruled out by inspection of the values at 4500-m depth, which show no corresponding increase along the path of flow.

The question of a possible contribution from the Amazon to the
bottom-water turbidity cannot be answered with the data at present.
The distribution of bottom values in the region north of the Amazon
from about 7° to 10°N suggests that bottom waters from the north, on
the western side of the basin, are higher in particulates than the
AABW from the south, whose core is located on the eastern side of
the basin [*Wright*, 1970]. The values of >0.7 log E/E_D in these latitudes are joined by contours to the North American Basin bottom
waters, but the paucity of data in the region of this joining should
be noted. Hence, it is uncertain whether the higher bottom values
in the region north of the Amazon are the result of particulates
gained at much higher latitudes or are the result of a more local
source, such as the Amazon River.

In the equatorial region, the bottom values range from .22 to
.68, with a majority of values between .4 and .6. The highest
values occur in the deeper parts and generally on the western side
of the basin. Ten stations in this region, all in depths less than
4000 m, showed no significant increase in turbidity — that is, the
minimum turbidity recorded was at the bottom.

DISCUSSION AND CONCLUSIONS

From the distribution of the background turbidity values associated with the intermediate-to-deep waters (Figure 2), it is
apparent that this background is a function of the rate of biogenic
fall-out from the surface waters. A pattern similar to the equatorial high observed here is also observed in the Pacific [*Lisitzin*,
1972 and *Thorndike*, in preparation, 1973]. In the western Atlantic,
the observed equatorial high in the background turbidity might also
have a terrigenous Amazon contribution in the region from 5° to 10°N.
The near-bottom velocities and associated shear of the Western
Boundary Undercurrent in the North American Basin [*Swallow and
Worthington*, 1961; *Volkmann*, 1962; and *Zimmerman*, 1971] and the
Antarctic Bottom Current in the Argentine Basin [*Wüst*, 1955; and
Le Pichon et al., 1971] produce high turbulent intensities which
allow the bottom waters to carry a load of suspended sediment in the
form of generally exponential decreases in suspended matter upward
in the bottom few hundred meters. This roughly exponential decrease
upward is seen both using nephelometers [*Eittreim and Ewing*, 1972;
Ewing et al., 1971] and methods that are more quantitative optically
such as measurements of the attenuation coefficient [*Matlack*, 1972].
It is also seen in measurements of particulate iron in bottom waters
of the North American Basin [*Betzer and Pilson*, 1971]. Such a turbidity profile is best explained by eddy diffusion away from a bottom boundary which acts as both a source and sink for particles and
by a steady supply of particles from upstream. The decrease in
bottom-water turbidity toward the equator from the north and south
may reflect both the increasing distance from particle sources and
decreasing bottom current velocities. It is difficult to separate
these two effects, since the bottom-water velocities tend to increase toward higher latitudes and since, at the high latitudes, the
North American and Argentine Basins are, perhaps, also source regions of particles, judging by the abundant submarine canyons along
their margins. It should be noted that the highest bottom turbidities were observed off the mouths of the deeply incised Ameghino and
Malvinas canyon systems of the southern Argentine continental slope
at 45°S to 48°S [*Lonardi and Ewing*, 1971], suggesting that these
canyons are a present-day source of particles to the nepheloid layer.
The turbidity profiles north of the Amazon River may suggest
some contribution from the Amazon River system to the deep-basin
sediment load. One high value (0.89 at bottom of 4598 m depth and
1.83 at 4500 m depth) in the basin due north of the Amazon River
mouth and somewhat higher values in this general region suggest
this. However, it is impossible to conclusively distinguish between
this source and that of southward flowing high-turbidity water from
the North American Basin on the basis of this data alone.

ACKNOWLEDGMENTS

F. Mouzo reduced part of the data used in this study; E. M. Thorndike was largely responsible for the data from cruise 15 and 16 of the R/V *Conrad*; L. Sullivan supervised the shipboard operations and data reduction; Mouzo, Thorndike, and Sullivan all provided helpful suggestions and discussions. P. Biscaye, Y. Lancelot, W. B. F. Ryan and D. Spencer critically reviewed the manuscript and their helpful comments are appreciated. The work was supported by the Office of Naval Research under Contract N-00014-67-0108-0004 and by the National Science Foundation under Grant GA27281. This paper is contribution 2033 of Lamont-Doherty Geological Observatory and contribution 30 of the Earth and Planetary Sciences Division, Marine Biomedical Institute, University of Texas Medical Branch.

REFERENCES

Be, A., and D. S. Tolderlund, Distribution and ecology of living planktonic Foraminifera in surface waters of the Atlantic and Indian Oceans, in *Micropaleontology of Oceans*, edited by B. M. Funnel and W. R. Riedel, pp. 105-149, Cambridge University Press, Cambridge, 1971.

Betzer, P. R., and M. E. Q. Pilson, Particulate iron and the nepheloid layer in the western North Atlantic, Carribean and Gulf of Mexico, *Deep Sea Res.*, *18*, 753, 1971.

Eittreim, S., P. M. Bruchhausen, and M. Ewing, Vertical distribution of turbidity in the South Indian and South Australian basins, in *Antarctic Oceanology II: The Australian-New Zealand Sector*, edited by D. Hayes, pp. 51-58, Amer. Geophys. Union, Washington, D. C. 1972.

Eittreim, S., and M. Ewing, Suspended particulate matter in the deep waters of the North American Basin, in *Studies in Physical Oceanography*, edited by A. L. Gordon, pp. 123-167, Gordon and Breach, London, 1972.

Eittreim, S., M. Ewing, and E. M. Thorndike, Suspended matter along the continental margin of the North American Basin, *Deep Sea Res.*, *16*, 613-624, 1969.

Ewing, M., S. Eittreim, J. Ewing, and X. LePichon, Sediment transport and distribution in the Argentine Basin: Part 3, Nepheloid layer and processes of sedimentation, in *Physics and Chemistry of the Earth*, vol. 8, pp. 49-77, Pergamon Press, New York, 1971.

Ewing, M., and E. Thorndike, Suspended matter in deep ocean water, *Science*, *147*, 1291-1294, 1965.

Kullenberg, G., Scattering of light by Sargasso Sea water, *Deep Sea Res.*, *15*, 423-432, 1968.

Le Pichon, X., S. Eittreim, and W. J. Ludwig, Sediment transport and distribution in the Argentine Basin; Part 1, Antarctic Bottom Current Passage through the Falkland Fracture Zone, in *Physics and Chemistry of the Earth,* vol. 8, pp. 1-28, Pergamon Press, New York, 1971.

Lisitzin, A. P., Sedimentation in the World Ocean, S.E.P.M. *Spec. Publ. 17,* 225 pp, 1972.

Lonardi, A. G., and M. Ewing, Sediment transport and distribution in the Argentine Basin: Part 4, Bathymetry of the continental margin, Argentine Basin and other related provinces, canyons, and sources of sediment, in *Physics and Chemistry of the Earth,* vol. 8, pp. 79-122; Pergamon Press, New York, 1971.

Matlack, D. E., Deep ocean optical measurements (DOOM), *Naval Ordnance Lab. Tech. Rept. No. 72-284,* 1972.

Owen, R. W., and B. Zeitzchel, Phytoplankton production: seasonal change in the oceanic eastern tropical Pacific, *Marine Biology, 1,* 32-36, 1970.

Swallow, J. C., and L. V. Worthington, An observation of a deep counter-current in the western North Atlantic, *Deep Sea Res., 8,* 1-9, 1961.

Thorndike, E. M., A deep-sea photographic nephelometer, in preparation, 1973.

Volkmann, G., Deep current observations in the western North Atlantic, *Deep Sea Res., 9,* 493-500, 1962.

Wright, W. R., Northward transport of Antarctic Bottom Water in the western Atlantic Ocean, *Deep Sea Res., 17,* 367-372, 1970.

Wüst, G., Stromgeschwindigkeiten in tiefen und Bodenwasser des Atlantischen Ozeans auf Grund dynmaischer Berechnung der Deutschen Atlantischen Expedition 1925-1927; Papers in Marine Biology and Oceanography, Suppl. *Deep Sea Res., 3,* 373-397, 1955.

Zimmerman, H. B., Bottom currents on the New England continental rise, *J. Geophys. Res., 76,* 5865-5876, 1971.

Variations in Benthic Boundary Layer Phenomena: Nepheloid Layer in the North American Basin

PIERRE E. BISCAYE AND STEPHEN L. EITTREIM

Lamont-Doherty Geological Observatory

ABSTRACT

Observations of phenomena associated with a benthic boundary layer were made over a 19-day period in two different regimes in the western North American Basin. Repeated measurements of temperature, in situ and in vitro turbidity, suspended particulate concentrations and excess radon versus depth, as well as bottom photographs, were made as a function of time at two locations on the lower slope of the Blake-Bahama Outer Ridge (BBOR) and at one location on the Hatteras Abyssal Plain (HAP). At the BBOR sites the benthic boundary layer was manifest by high concentrations of suspended particulates, high turbidity, and intense vertical mixing indicated by excess radon. Vertical distributions of particulate matter and radon were related to the thermal structure of the water mass and, although the strongest manifestations of frictional interaction between the water and bottom were seen below 150-300 m, some influence was seen as high as 1500 m above bottom.

The 150-300-m particulate and radon boundary also coincided with the top of an adiabatic layer. Large, regular, temporal variations in these parameters with a period of about one week were in phase at the two stations, located 110 km apart. This benthic boundary layer regime was associated with current velocities from 10 to 30 cm/sec and with a current direction and temperature and salinity characteristics indicative of the Western Boundary Undercurrent (WBUC). At the HAP site, manifestations of the benthic boundary layer were less intense, less variable and restricted to a thinner layer. Turbidity and concentration of particulate matter were much lower than on the BBOR and vertical mixing measured by excess radon was an order of magnitude lower. Depth profiles of these parameters showed the strongest evidence of the benthic

*boundary layer to be restricted to 80-100 m above the bottom
(cf. 150-300 m on the BBOR) but with some evidence of mixing of
particulate matter and cold bottom waters to 900 m above the
bottom (cf. 1500 m on the BBOR). Again, the top of the 80-100-m
zone coincided with the top of the adiabatic bottom layer. Bottom
photographs and water mass characteristics indicate that this
benthic boundary layer regime is associated with the rather slowly
northward moving Antarctic Bottom Water (AABW). An hypothesis to
explain the variations observed on the BBOR is presented in which
eddies of clear AABW are injected into the more turbid, rapidly
southward flowing WBUC north of the study area. These are seen as
temporal variations in the intensity of the nepheloid layer along
the BBOR.*

INTRODUCTION

Adjacent to the bottom of the world oceans, a zone exists in
which frictional forces result from the interaction of the deep
oceanic circulation with the ocean bottom. This zone of frictional
interaction is identified here as the benthic boundary layer as it
was by *Wimbush and Munk*[1970]. The latter review theoretical
considerations applicable to the zone as well as some direct
measurements. The phenomena discussed in the present study extend
to heights above the bottom, orders of magnitude greater than those
discussed by Wimbush and Munk. The difficulty of observing the
benthic boundary layer in the deep ocean is a principal reason that
the number of measurements is meager. The measurements reported
by Wimbush and Munk are of the density and velocity fields and
the benthic boundary layer they observed off southern California
had a thickness of meters. *Sternberg* [1970] also made velocity
profile measurements in the benthic boundary layer using a dye
pulse injection device; his observations also were restricted to
meters above the bottom.

By far the largest number of direct observations of phenomena
associated with the benthic boundary layer consist of measurements
of suspended particulate matter that constitutes the nepheloid
layer. The nature and intensity of this layer of relatively
turbid water have been observed most extensively over the world
oceans (principally by the Lamont-Doherty group) by *in situ*
measurements of light scattering [*Ewing and Thorndike*,1965;
Hunkins et al., 1969; *Eittreim et al.*, 1969, 1972a, b; *Ewing and
Connary*, 1970,*Ewing et al.*, 1971; *Eittreim and Ewing*,1972;
Connary and Ewing, 1972]. Associated with most of the approximately
3000 Lamont-Doherty nephelometer stations have been bottom photographs, usually about 10 per station. Although not direct measurements of the benthic boundary layer, the nephelometer profiles
demonstrate the spatial variability of frictional forces in the
benthic boundary layer and bottom photographs often show effects
of these frictional forces in eroding and redistributing bottom
sediment. The nephelometer profiles constitute data instantaneously
observed in the benthic boundary layer, whereas the bottom photographs

show time-integrating phenomena such as ripple marks or scour
marks which may enter the geologic record by subsequent burial.

At about the same time that measurements of light scattering
were begun at Lamont, *Broecker* and co-workers [1965; *Broecker et
al.*, 1967; *Broecker et al.*, 1968] demonstrated the utility of the
naturally occurring, noble-gas, radioactive tracer radon-222 in
studying near-bottom mixing. In this method, the radon which
diffuses out of the sediments (called excess radon) is extracted
from a profile of samples spaced from tens to hundreds of meters
above the bottom. The vertical gradient of this radon concentration may be used to calculate vertical mixing parameters.

As originally envisioned by Broecker and his co-workers,
a regular exponential decrease of radon concentration above the
bottom could be used to characterize vertical mixing or eddy
diffusibility. They found a number of vertical distributions of
excess radon from both Atlantic and Pacific to be sufficiently
close to an exponential case to suggest the general utility
of an exponential model. In attempts to understand the mechanisms
responsible for the nepheloid layer, Biscaye measured the near-
bottom vertical distribution of excess radon and suspended particulate matter of water samples on transects across the western
North Atlantic and the South Atlantic Oceans. The results
indicated no apparent first-order correlation between eddy
diffusivity (where measureable) and the concentration of particulate
matter suspended in the same water in any one cast. The radon
profiles were also sufficiently complex to suggest that the
"classical" exponential distributions with height were less
common than originally supposed, a suggestion subsequently rein-
forced by the radon profiles of *Chung and Craig* [1972] in the
Pacific Ocean. The lack of correlation between gross particulate
distributions and radon-mixing parameters suggest that, while
they may respond to similar mixing stimuli, they have different
relaxation times: the radon being dependent upon its decay
constant and the particulate matter being dependent on its range
of settling velocities. In the last several years, Gad Assaf of the
Weizmann Institute, Israel, has worked with the Lamont-Doherty
group on using the excess-radon method and recognized its utility
in defining the benthic boundary layer.

The several methods mentioned for observing the benthic
boundary layer (current velocity, light scattering and excess-radon
profiles) suggest widely varying dimensions and mixing intensities
as a function both of geography and the observational method.
The greatest number of measurements, those made by light scattering,
show a sufficiently coherent large-scale geographic pattern to
suggest a steady-state process of years to decades. However, very
few simultaneous measurements have been made of even two of the
phenomena noted, much less several or all of them. To our knowledge, no attempts have been made to eliminate the geographic variable or to observe the degree to which these phenomena are steady-
state on some time scale. On the one or two occasions on which
casts for near-bottom observations or sampling were repeated on
a time scale of hours or days by either of the present authors,

significant variability in near-bottom light scattering, suspended-particulate profiles, and excess-radon profiles was suggested. However, these occasions were not specifically designed to eliminate the geographic variable, and the drift of the ship between casts left open the question of whether the variability was geographic or temporal.

PLAN OF THE EXPERIMENT

In order to test the short-term variability of phenomena associated with the benthic boundary layer, an experiment was designed in which sampling and observations were repeated over a period of days and geographic variability was minimized by means of satellite navigation. This program was carried out on R/V *Robert D. Conrad* in August and September 1972 in the western North Atlantic Ocean. Three locations were chosen to reflect the variability under two different near-bottom regimes. Two locations separated by 110 km, on the lower slope of the Blake-Bahama Outer Ridge (BBOR) were chosen to observe the variability under the current regime of the Western Boundary Undercurrent (WBUC). The WBUC is a western boundary bottom current of high, north-latitude sources [*Swallow and Worthington*, 1961; *Lee and Ellet*, 1967] and perhaps some admixture of Antarctic Bottom Water (AABW) [*Amos et al.*, 1971] with which a strong nepheloid layer has been identified [*Eittreim et al.*, 1969]. Stations E and F (Figure 1) along the strike of the bathymetric contours represented locations in which some small-scale local topographic relief (meters to tens-of-meters) interacts with a fairly strong, south-flowing bottom-current regime. The BBOR is entirely a sedimentary feature: seismic reflection profiling shows an essentially flat underlying basement. It is a large, oceanic, clay/silt dune or drift deposit jutting southeastward from the North American continental slope. It represents a long-term depositional sedimentary regime, whether or not deposition or erosion is presently occurring.

Station G (Figure 1) on the Hatteras Abyssal Plain (HAP) was chosen to provide significantly different conditions. A cold water mass showing remnant characteristics of the Antarctic Bottom Water (AABW) was anticipated in the benthic boundary layer regime at this station. From previous nephelometer stations the intensity and thickness of the nepheloid layer was anticipated to be less than that at stations E and F. The sedimentary regime at station G represents abyssal-plain (probably turbidity-current) deposition on a geologic time scale, although, on the time scale of this experiment it is unknown. A further contrast with stations E and F is the complete lack of local topographic relief greater than centimeters to a meter.

The experiment was conducted in the following manner. The first ten days were spent shuttling back and forth between stations E and F, making at least one set of casts per day at each station. Attempts were made to time the casts so that samples and near-bottom

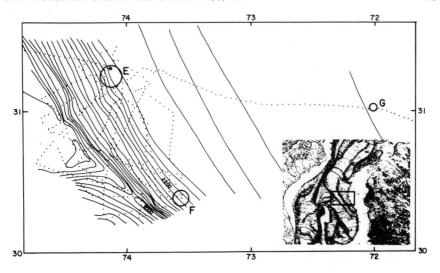

Fig. 1. Location of stations E, F, and G and bathymetry in
nominal fathoms from soundings obtained during between-stations
steaming (dotted lines = ship's track). A portion of the crest of
the Blake-Bahama Outer Ridge is defined by the 1800 to 2000-m
contours. Circles = areas of ship locations when casts reached
bottom, except for station 5 (located northeast of circle at
station E), which was affected by extremely high eastward surface
currents. * = location of current meter at station E. A portion
of the physiographic diagram of the North Atlantic [*Heezen and
Tharp*, 1968] shows the general physiographic setting of the study
area.

in situ measurements were made at the same relative time on the
semidiurnal tidal cycle. Tracks between stations E and F varied,
chosen to yield maximum seismic profiling coverage of the BBOR.
After ten days at stations E and F, station G was occupied seven
times in the succeeding six days and finally, station F was
reoccupied three times during the last two days of station time.
The list of stations, positions, and casts for the 19-day
experiment is given in Table 1.
    The experiment was designed to eliminate geographic variability.
Whereas satellite navigation permits knowledge of the ship's
position to within several hundred meters, obviously it does not
control the ship's location or the position of a sampling/instru-
ment package five kilometers below the surface. Strong surface
currents (up to 4.5 knots), therefore, turned out to be a
significant problem in reoccupying the exact locations of stations,
particularly station E, station F to a lesser degree, and station
G to a much lesser degree. This is seen in the relative size of the
circles indicating location areas of stations E, F, and G in Figure
1. The degree of geographic variability experienced in reoccupying
the stations did not affect the conclusions drawn from this
experiment, as will be shown.

TABLE 1.   List of Stations, Positions, and Casts (RC 16-1).

Ship Station	Date 1973	Cast*	Time	Lat(N)	Long(W)	Water Depth	Ship Station	Date 1973	Cast*	Time	Lat(N)	Long(W)	Water Depth
				At Bottom or Trip							At Bottom or Trip		
3 E	18,19 Aug	STD1,Rn1	1250	31-16.2	74-04.5	5021	13 E	28 Aug	STD11	1020	31-15.5	74-13.3	4900
		N2	1748	31-12.5	74-05.1	4945			N11,Rn13	1437	31-20.7	74-09.6	4895
		STD2,Rn2	0107	31-14.9	74-04.7	4983	14 G	29 Aug	STD12,Rn14	1221	31-01.1	72-01.0	5404
4 F	19 Aug	N3	1037	30-22.6	73-34.2	4821			N12	1604	31-00.9	71-58.8	5404
		STD3,Rn3	2222	30-22.6	73-37.7	4679	15 G	30 Aug	N13,Rn15	1354	31-02.1	72-00.1	5404
5 E	20 Aug	STD4,Rn4	1401	31-16.7	73-57.5	5089	16 G	31 Aug	N14,Rn16	1441	31-02.0	72-01.0	5404
		N4	1803	31-15.8	73-58.0	5073	17 G	1 Sept	N15,Rn17	0742	31-02.2	72-01.6	5404
6 F	21 Aug	STD5,Rn5	1539	30-23.2	73-33.8	4803			N16,Rn18	1505	31-02.4	71-58.0	5404
		N5	1900	30-22.5	73-36.8	4685			STD14	1935	31-01.3	71-58.9	5404
7 E	22 Aug	STD6,Rn6	1537	31-19.0	74-08.0	4990	18 G	2 Sept	N17	0941	31-01.7	72-00.8	5404
		N6	1926	31-20.2	74-07.3	4962			STD15,Rn19	1540	31-02.8	71-58.3	5404
8 F	23 Aug	Rn7	1655	30-26.5	73-36.7	4805	19 G	3 Sept	N18	0817	31-01.1	72-01.3	5404
9 E	24 Aug	Rn8,N7	1115	31-16.4	74-13.0	4800			STD16,Rn20	1310	31-01.9	72-01.4	5404
		STD7,Rn9	1830	31-17.7	74-09.1	4889	20 F	4 Sept	N19	0713	30-23.5	73-35.4	4750
10 F	25 Aug	N8	1324	30-25.5	73-33.5	4871			STD17,Rn21	1508	30-20.8	73-33.0	4760
		STD8,Rn10	1857	30-24.3	73-37.7	4708	21 F	5 Sept	STD18,Rn22	1211	30-22.3	73-36.6	4675
11 E	26 Aug	N9	1232	31-17.9	74-11.2	4822			N20	1452	30-22.1	73-37.1	4658
		STD9,Rn11	1842	31-21.0	74-08.4	4968	22 F	6 Sept	STD19,Rn23	1200	30-22.5	73-36.2	4760
12 F	27 Aug	N10	1427	30-24.4	73-37.2	4749			N21	1450	30-22.6	73-37.2	4660
		STD10,Rn12	1927	30-24.0	73-38.2	4690							

*STD = Salinity, Temperature, Depth   Rn=Radon   N=Nephelometer-camera

SAMPLING, MEASUREMENTS, AND ANALYTICAL RESULTS

Current meter measurements were made continuously at station E. Bottom photography and *in situ* measurements were made at each occupation of each station for salinity, temperature, depth (STD), and turbidity. Water samples were taken for suspended particulate matter, *in vitro* turbidity, and for measurements of radon concentration. Descriptions of these measurements follow.

A Geodyne-brand, pop-up Richardson-type current meter was dropped at the first occupation of station E and recorded data for 10.5 days during reoccupations of stations E and F. It was tethered one meter above the bottom and recorded current magnitude and direction every 15 seconds. The same current meter was dropped at the first occupation of station G and was recovered six days later but no data were obtained due to flooding resulting from a ruptured O-ring.

The Bissett-Berman Model 9006 STD system continuously measured salinity and temperature versus depth and is calibrated by sampling with a deck-activated 1.7-ℓ Niskin-bottle array fitted with reversing thermometers. The salinity sensor experienced considerable drift and the on-deck salinometer (Bissett-Berman Model 6230) functioned properly only intermittently; water-mass characteristics are based principally on temperature measurements. The temperature traces are shown in Figure 2. Note the adiabatic gradients in the bottom layers at the three deep HAP stations (G) and at most of the other stations (E and F). Salinity measurements are given in Table 2. Comparison of salinities from the 30-ℓ bottles with those from the 1.7-ℓ bottles suggests that the 1.7-ℓ bottles leaked during ascent from the bottom. The low accuracy of the salinity measurements combined with the small differences likely

TABLE 2. Salinity Measurements on Bottom Bottle Samples

Station	STD Cast	Salinity*	Radon Cast	Salinity**
E	STD1	34.899	Rn1	34.872
E	STD2	34.888	Rn2	34.869
F	STD3	34.921	Rn3	34.951
E	STD4	34.895	Rn4	34.876
F	STD5	34.988	Rn5	——
E	STD6	——	Rn6	34.883
F	——	——	Rn7	34.887
E	——	——	Rn8	34.905
E	STD7	34.937	Rn9	34.884
F	STD8	34.903	Rn10	34.891
E	STD9	34.885	Rn11	——
F	STD10	34.900	Rn12	34.891
E	STD11	34.882	Rn13	34.878
G	STD12	34.877	Rn14	34.889
G	——	——	Rn15	——
G	STD13	——	Rn16	34.863
G	——	——	Rn17	34.878
G	STD14	——	Rn18	34.870
G	STD15	——	Rn19	34.857
G	STD16	——	Rn20	34.877
F	STD17	——	Rn21	34.902
F	STD18	——	Rn22	34.881
F	STD19	——	Rn23	34.885

Mean	Stations		
	E	F	G
1.7-ℓ Samples	34.900	34.908	-
30-ℓ Samples	34.881	34.898	34.872

*using 1.7-ℓ Niskin bottles
**using 30-ℓ Niskin bottles

to be found in salinity between lower North Atlantic Deep Water (NADW) and AABW at this latitude [*Amos et al.*, 1971] reduce the significance of the salinity measurements except to show that salinities measured were those to be expected for the water masses encountered based on the literature [*Fuglister*, 1960].

A continuous *in situ* record of turbidity versus depth was measured at each station using a standard Lamont photographic nephelo-·meter [*Thorndike and Ewing*, 1967; *Eittreim et al.*, 1969; *Thorndike*, in preparation, 1974]. Mounted on the same frame as the bottom camera, this instrument consists of a 35-mm camera with continuous film drive, an opposing light source at a distance of 55 cm and, midway between these, an intervening baffle-attenuator which cuts

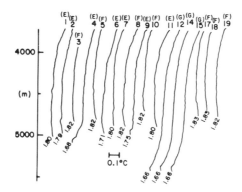

Fig. 2. Lower portion of *in situ* temperature profiles obtained using STD system (precision ±0.02°). Potential temperatures, calculated from the STD output, corrected using reversing thermometer data, are indicated for bottom-most data points.

off scattering from angles <7 degrees. A direct beam of the source light is allowed to pass to the camera, attenuated by a factor of about 10^3 by a combination of an opal glass and a gray filter whose transmittance is measured and recorded as a constant of the instrument. Depth is recorded on the film by deflection of a light spot through a bourdon-tube pressure sensing system. The units of measure are a ratio of the film exposures produced by the scattered light E to that produced by the direct, attenuated light E_D. The use of the ratio E/E_D compensates for any changes in intensity of the light source, transport speed of the film, developing of the film, or film sensitivity and is a function of scattering only, produced over the range of angles from approximately 7 to 30 degrees. The rate of film movement, the width of the camera slit and the lowering rate of the nephelometer combine to integrate the scattering over 20 m of the water column. Nephelometer profiles and about 12 bottom photographs were made at each station occupation, with a total of 13 sets at stations E and F and 6 at station G. The nephelometer profiles are shown in Figures 3 and 4. Note, in Figure 3, the greater intensity of scattering and the greater variability at stations E and F. Note, also, that the nepheloid layer extends higher in the water column at stations E and F than at station G. In the bottom portions of the nephelometer profiles (Figure 4), note that the most intense light scattering occurs below about 300 m on the BBOR. For the stations at F, note the close agreement between the top of the adiabatic layer and high light-scattering gradients.

Water samples were taken primarily using eight 30-ℓ Niskin bottles, although *in vitro* turbidity and salinity measurements were also made on water from the 1.7-ℓ Niskin bottles using the STD system. The eight 30-ℓ bottles were clamped on the lower 100 to 400 m of wire above either the STD system or the nephelometer camera. A

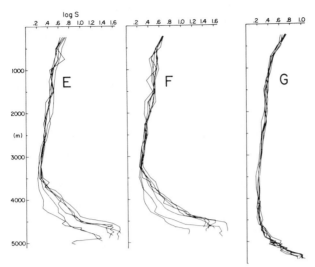

Fig. 3. All nephelometer profiles at stations E, F and G, superimposed. $S = E/E_D$, the ratio of film exposure on the scattering trace to that on the trace recording attenuated direct light.

pinger was used to determine height above the bottom, enabling the instrument to be held at a known number of meters above the bottom until arrival of the messenger to trip the bottles, taking samples at precisely known intervals.

Sampling for radon and suspended particulate matter was accomplished by sucking about 19 ℓ of water from the Niskin bottle, through a filter held in an in-line filter holder and into a previously evacuated 20-ℓ flint-glass bottle. The filters used were 47 mm diameter, 0.4 μm Nuclepore-brand filters that had been weighed in the laboratory under controlled temperature and humidity. These filters retain particles larger than 0.4 μm and a large fraction of the particles smaller than 0.4 μm by impaction of the particles on the filter. Some workers feel that the equivalent Millipore-brand filters (HA type, 0.45 μm) are more efficient than Nuclepore-brand at retaining particles smaller than the nominal pore diameter. Nuclepore filters were used in this application for the following reasons. (1) Nuclepore filters weigh less than an equivalent Millipore filter (∼ 20 mg versus ∼ 90 mg) so that, in obtaining the weight of suspended particulate matter from a 20-ℓ sample of clean water (containing perhaps tens of micrograms), a small-difference measurement is made on a smaller total weight. (2) Millipore filters are coated with a wetting agent which washes out upon use, creating a weight difference of the order of the suspended particulate matter burden. We have not detected such a weight loss in Nuclepore filters. (3) Due to their lower surface area and possibly the lack of a wetting agent, Nuclepore filters achieve relative humidity equilibrium much more rapidly and suffer much less change in weight due to change in weighing-room relative humidity, reducing (but not

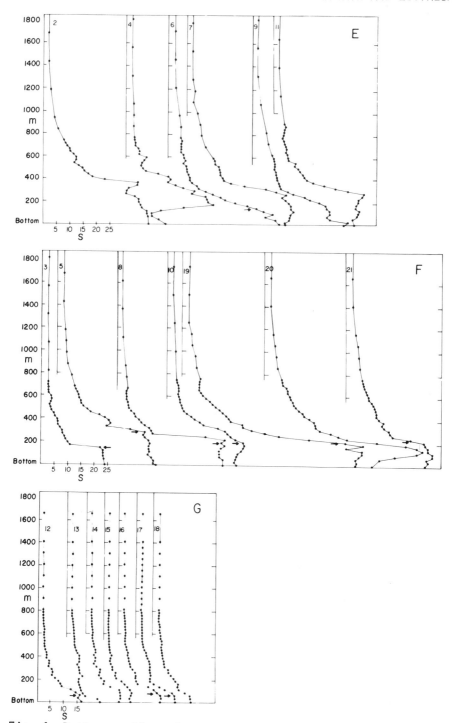

Fig. 4. Bottom portion of nephelometer profiles at stations E, F and G. Arrows = top of adiabatic layers where they are defined by well-delineated temperature minima.

eliminating) the degree of relative-humidity control required in the weighing-room or weighing chamber. (4) Because of the difference in the nature of the filter, the surface of Millipore filters have enormously more visual "texture" than do Nuclepore filters. For scanning electron microscopy work, therefore, at a given magnification one can distinguish the presence of small particles much more readily on a Nuclepore filter. (5) If any analytical work on trace elements is to be done on the samples, Nuclepore filters have a significantly lower blank for a number of elements than do the Millipore Type HA filters (Derek Spencer, personal communication).

As soon as possible after filtration, the filters were washed with double-distilled, filtered water to remove salt. All operations using the filters in the lab and aboard ship were done in a laminar-horizontal-flow (LHF) hood which provides a continuous flow of ultra-filtered air. The filter holder was attached to a suction flask in the LHF hood and the filter washed with fourteen aliquots of water (totalling about 50 mℓ). This washing procedure was checked for complete salt removal using silver nitrate and the Hach turbidimeter (described below). Results of the gravimetric analysis are reported in Table 3. The precision of particulate concentration analyses is estimated at ±5% for concentrations >20µg/ℓ, increasing to ±25% for concentrations to 4µg/ℓ. The most significant error for concentrations >20µg/ℓ results from the measurement of water volume using the calibrated flint-glass bottles used for radon extraction.

The filtered sea water was held in a 20-ℓ flint-glass bottle for extraction of dissolved radon. Because of the short half-life of radon (3.85 days) this must be done as soon as possible. The extraction procedure requires less than two hours.

The radon boards used on the R/V *Conrad* utilized the principle of *Broecher* [1965] although they were materially improved in valving, tubing and layout from the system he describes. For the extraction of dissolved radon, helium is bubbled through the sea water, passed through a liquid nitrogen trap where radon, CO_2 and H_2O are condensed and the helium is recycled through the sea water. After 90 minutes the trap is thawed, and the radon is purified by circulating through Ascarite and $Mg(ClO_4)_2$ which quantitatively remove CO_2 and water. The radon is again trapped in a small volume at the temperature of liquid nitrogen and then expanded and flushed into an evacuated scintillation cell. The cell is removed from the radon board and the sample is counted. Counting times aboard ship varied with sample activity and the pressure of other samples to be counted, but in general more than 1000 counts were acquired as a minimum. The overall error on analysis for excess radon is ∿ 10%. The results are given in Table 3.

The analysis on shipboard of near-bottom sea-water samples for radon includes both "excess" radon, or that which diffuses across the sediment/water interface, and "supported" radon or that existing in sea water in radioactive equilibrium with its radium-226 parent which is dissolved in sea water. It is because the concentration of radium-226 and, therefore, its radon daughter is much higher in sediments than in sea water that a strong concentration gradient exists across the sediment/water interface, providing the

TABLE 3. Concentrations of Excess Radon and Suspended Particulate Matter, and *in vitro* Turbidity (in Formazin Turbidity Units) for Stations E, F, and G.

Cast	Station	Meters Above Bottom	Excess Radon* (dpm/100 l)	Suspended Solids (ug/l)	FTU
Rn1	E	106	6.8	178	0.17
		44	5.5	174	0.17
		33	6.9	183	0.19
		24	6.5	209	0.22
		15	.	200	0.19
		9	6.7	168	0.23
		6	4.6	166	0.28
Standing Crop			1.65 dpm/cm²	4.52 mg/cm²	
Rn2	E	106	8.6	188	0.21
		63	6.8	94.6	0.23
		43	4.8	227	0.19
		32	1.2	198	0.24
		24	5.5	181	0.20
		14	3.6	188	0.23
		9	5.6	169	0.24
		5	7.0	182	0.27
Standing Crop			1.91 dpm/cm²	4.52 mg/cm²	
Rn3	F	200	0.7	80.6	0.13
		102	2.1	118	0.12
		82	2.2	83.9	0.13
		62	1.9	130	0.14
		42	8.8	143	0.13
		22	5.7	145	0.14
		13	6.3	131	0.15
		6	5.0	127	0.16
Standing Crop			0.67 dpm/cm²	3.04 mg/cm²	
Rn4	E	291	0.9	23.2	0.06
		200	-0.2	20.8	0.05
		103	-1.2	27.7	0.06
		75	5.4	22.4	0.09
		57	3.3	26.7	0.06
		29	12.1	23.7	0.06
		20	6.2	30.9	0.07
		6	8.0	24.9	0.06
Standing Crop			0.66 dpm/cm²	0.82 mg/cm²	
Rn5	F	358	.	0.0	0.05
		202	4.0	43.3	0.10
		111	-0.1	29.9	0.10
		74	6.0	34.2	0.10
		56	2.0	72.3	0.10
		37	1.7	30.2	0.10
		19	4.8	30.9	0.11
		5	.	40.4	0.11
Standing Crop			0.93 dpm/cm²	1.11 mg/cm²	
Rn6	E	408	-3.0	10.8	0.05
		289	-2.0	10.0	0.05
		207	-2.5	20.5	0.06
		115	2.9	63.2	0.09
		66	2.8	113	0.11
		46	2.5	94.3	0.12
		26	4.1	111	0.12
		9	0.8	86.5	0.12
Standing Crop			0.37 dpm/cm²	1.72 mg/cm²	
Rn7	F	408	.	7.65	0.04
		262	-3.0	90.4	0.14
		161	1.0	119	0.20
		115	5.0	157	0.28
		70	2.8	192	0.29
		42	2.2	183	0.30
		24	4.4	180	0.39
		9	6.4	684	0.83
Standing Crop			0.60 dpm/cm²	5.31 mg/cm²	
Rn8	E	351	-2.3	20.5	0.07
		252	-2.4	120	0.17
		201	.	105	0.19
		152	-1.0	143	0.20
		102	4.0	144	0.24
		51	3.4	99.5	0.22
		27	5.0	128	0.26
		7	7.6	135	0.25
Standing Crop			0.55 dpm/cm²	3.89 mg/cm²	
Rn9	E	357	-3.4	13.7	0.11
		286	-1.0	55.1	0.15
		205	-3.3	182	0.35
		146	0.0	176	0.34
		106	1.0	193	0.36
		86	8.6	185	0.37
		55	7.7	145	0.32
		9	6.0	150	0.33
Standing Crop			0.74 dpm/cm²	4.73 mg/cm²	
Rn10	F	355	-2.4	17.3	0.06
		300	0.5	37.1	0.07
		208	.	170	0.24
		154	2.8	170	0.27
		108	0.2	178	0.27
		84	5.6	151	0.30
		44	2.8	174	0.31
		11	6.6	152	0.33
Standing Crop			0.78 dpm/cm²	4.61 mg/cm²	
Rn11	E	Cast mistripped at an unknown intermediate depth			
Rn12	F	353	-0.3	29.3	0.09
		252	0.6	62.8	0.11
		207	0.4	67.4	0.16
		152	-2.4	118	0.20
		106	-0.8	119	0.20
		79	3.8	95.7	0.18
		42	3.7	76.7	0.18
		9	8.2	115	0.17
Standing Crop			0.49 dpm/cm²	2.91 mg/cm²	
Rn13	E	284	-1.8	50.0	0.13
		210	0.7	71.6	0.16
		156	-2.4	69.5	0.17
		128	-2.6	66.1	0.16
		110	2.0	73.3	0.16
		92	0.5	79.5	0.16
		48	5.4	94.6	0.17
		13	5.4	73.9	0.17
Standing Crop			0.43 dpm/cm²	2.35 mg/cm²	
Rn14	G	304	0.4	2.63	0.05
		205	3.8	3.22	0.04
		104	1.1	5.15	0.06
		84	0.9	10.3	0.06
		44	9.4	18.2	0.07
		33	25	29.2	0.06
		18	12	16.1	0.07
		8	21	21.8	0.07
Standing Crop			1.50 dpm/cm²	0.25 mg/cm²	
Rn15	G	293	-2.6	7.14	0.06
		205	2.3	9.18	0.05
		150	1.3	16.3	0.04
		95	0.5	13.7	0.04
Bottom four bottles did not trip					
Rn16	G	353	-2.4	0.00	0.04
		254	0.8	0.00	0.04
		154	.	3.68	0.05
		104	.	15.7	0.07
		79	-0.2	15.7	0.07
		53	2.7	15.5	0.07
		33	5.6	12.6	0.07
		7	19	12.3	0.07
Standing Crop			0.64 dpm/cm²	0.22 mg/cm²	
Rn17	G	354	-3.8	1.67	0.04
		255	0.4	7.22	0.04
		155	0.1	2.70	0.05
		105	1.2	29.7	0.06
		80	.	8.82	0.06
		50	-0.5	17.3	0.06
		34	13	25.9	0.07
		8	34	26.2	0.07
Standing Crop			1.07 dpm/cm²	0.39 mg/cm²	
Rn18	G	306	.	4.74	0.04
		205	0.4	9.28	0.05
		134	5.1	16.0	0.05
		84	-0.4	23.2	0.06
		71	1.3	21.5	0.06
		27	8.4	32.5	0.06
		18	13	32.6	0.06
		7	27	66.2	0.14
Standing Crop			1.08 dpm/cm²	0.56 mg/cm²	
Rn19	G	306	0.0	6.01	0.05
		206	-0.9	7.49	0.05
		132	-1.2	0.52	0.05
		86	-0.2	21.5	0.06
		50	4.8	23.2	0.07
		31	20	23.1	0.07
		18	23	17.2	0.07
		9	23	19.3	0.07
Standing Crop			1.01 dpm/cm²	0.36 mg/cm²	
Rn20	G	355	-2.8	5.89	0.05
		208	-2.0	9.19	0.05
		104	0.1	12.3	0.06
		84	1.1	12.4	0.06
		64	2.8	18.9	0.06
		44	7.3	7.89	0.07
		24	16	14.4	0.06
		11	26	13.3	0.07
Standing Crop			0.95 dpm/cm²	0.37 mg/cm²	
Rn21	F	394	0.4	17.2	0.06
		246	-2.3	162	0.20
		125	1.0	191	0.23
		104	5.3	184	0.25
		84	3.6	171	0.24
		55	-0.4	164	0.24
		24	5.5	181	0.26
		6	9.5	136	0.25
Standing Crop			0.50 dpm/cm²	5.61 mg/cm²	
Rn22	F	400	0.1	15.7	0.07
		251	-0.3	70.0	0.12
		126	-3.3	150	0.20
		104	-0.7	136	0.22
		83	0.8	126	0.22
		53	3.1	139	0.21
		22	10	61.7	0.17
		7	14	88.1	0.17
Standing Crop			0.55 dpm/cm²	3.48 mg/cm²	
Rn23	F	450	-	7.5	0.12
		254	-	47.4	0.20
		128	-	122	0.23
		108	-	112	0.27
		84	-	72.3	0.23
		55	-	100	0.17
		26	-	76.8	0.18
		11	-	89.5	0.17
Standing Crop			-	2.77 mg/cm²	

*Excess radon above 15.1 dpm/100 l radon supported by dissolved radium-226.

bottom-source tracer for "excess" radon utilized in measuring
vertical mixing. Excess radon is thus the difference between the
total (shipboard-analyzed) radon concentration and that of dissolved radium in the sea water. In the laboratory, measurement of
radium was made from samples of water from a number of casts. The
measurement procedure is exactly the same as above because,
once the radon daughter grows into equilibrium with its parent
radium, their concentrations are the same and the radium can be
measured by extracting and measuring the radon. Results of the
radium measurements made in the laboratory are given in Table 4.
A weighted average of all the measured radium concentrations of

TABLE 4.   Dissolved Radium

Cast	Station	Meters Above Bottom	Meters Water Depth	Radium dpm/100 ℓ	σ*
Rn1	E	106	4904	15.4	2.9
		44	4966	15.5	2.2
Rn3	F	42	4635	15.3	0.12
		6	4671	15.3	1.5
Rn6	E	408	4573	14.1	0.67
		115	4866	14.6	0.37
		9	4972	14.6	0.48
Rn9	E	205	4679	15.3	1.3
		55	4829	15.0	0.39
Rn10	F	208	4685	15.4	1.0
Rn18	G	7	5393	15.3	0.42

*The radium concentration is the average of at least two
analyses per sample and the σ given is based on the range of those
results.

15.1 dpm/100 ℓ has been subtracted from all total-radon concentrations
made on shipboard, irrespective of station, to yield excess radon.
Use of this single radium value may contribute to some of the
scatter in the excess-radon data. The particularly high scatter in
the water column may be reduced by use of individual radium
analyses for each sample, for each cast or even for each station
(E, F, and G) but insufficient analyses exist to justify even
distinguishing different radium concentrations for each station.
There may, therefore, be a constant bias of about 1 dpm/100 ℓ
for the excess-radon data. It should thus be noted that the
radium subtraction critically affects the height above the bottom

to which it can be concluded that excess radon has been mixed,
particularly for the low-radon concentrations in the North
Atlantic Ocean. This, of course, affects eddy diffusivities
calculated from the excess-radon data. The conclusions in this
report drawn from the radon data are, however, based upon large
differences and are not affected by any bias that may exist in
the radium value used.

Shipboard measurements for turbidity were also made on
sampled water using a Hach turbidimeter (Model 2100 A). Like the
nephelometer, this instrument measures light scattering from
particles in the water but measures at 90° angles to the light
beam rather than the low-angle scattering measurement of the
nephelometer. The utility of the turbidimeter is primarily in
providing an almost immediate indication of near-bottom turbidity,
thereby aiding in planning the spacing of the 30-ℓ Niskin bottles
on the wire for the next cast. Figure 5 shows the correlation
between turbidity as measured *in vitro* using the turbidimeter and
suspended particulate concentration as measured in the laboratory.
Although the scatter in the data is appreciable, the correlation
coefficient is 0.87, confirming the utility of the turbidimeter on
shipboard.

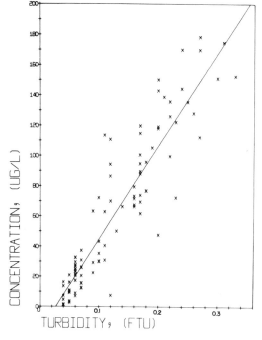

Fig. 5. Turbidity in Formazin Turbidity Units, measured *in
vitro* using the shipboard turbidimeter, versus concentration of
particulate matter (µg/ℓ), measured on filtrates from the same
samples.

The advantage of using a nephelometer or other *in situ*
measuring device is that a continuous record with depth is obtained,

limited in vertical resolution only by the relative rates of data acquisition and instrument lowering. In addition to providing indications of relative light-scattering intensity this continuous record becomes much more valuable if it can be translated into geologically — or geochemically — significant units such as volume or mass concentration. The scattering values S recorded by the nephelometer have been plotted against concentrations of particulate matter in Figure 6. A linear regression analysis

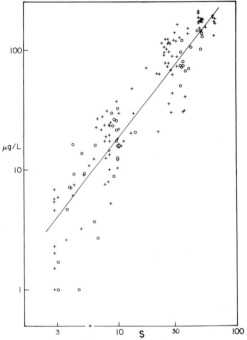

Fig. 6. Concentration of particulate matter from water samples versus *in situ* scattering value recorded using the nephelometer at sample depth. Circles = data from samples taken on the same wire as the nephelometer; crosses = data from casts taken in sequence, the nephelometer cast lagging the radon/particulate cast by from three to five hours, and in one case by eight hours.

gives a correlation coefficient of 0.93, indicating that the primary cause of the scattering variations is variations in particulate concentration. What the levels of particulate concentrations are in the clearest intermediate waters remains unanswered from these data. Certainly it is less than 10 µg/ℓ but limits of precision in the filter-weight determinations preclude specifying the value much closer than that. This determination awaits the collection of larger uncontaminated water samples for filtration.

Another type of sampling, piston coring, was regularly done during this experiment simultaneously with each nephelometer/camera station. Results on these samples will be reported in a future communication.

DISCUSSION

Temperature Profiles

Only temperature/depth information was obtained with the STD system due to the frequent failure of the salinity sensor, already noted. On each lowering, reversing thermometers were tripped for calibration of the STD temperature output. The difference between thermometer data and the raw STD temperatures was .02° C. Potential temperatures for the bottom-most data point are indicated below each curve in Figure 2. Temperatures consistent with an AABW source were observed in the HAP [Amos et al., 1971]. The BBOR stations showed warmer potential temperatures by about 0.1° C or more. An adiabatic, cold, bottom layer with a well-defined "lid" 60 to 80 m above bottom was consistently observed at the HAP site. These "lids" are zones of high thermal gradient and are similar to those observed by Amos et al. [1971]. They would be expected to form at the top of a well-mixed boundary layer in the presence of a strong negative temperature gradient such as exists in the bottom 500 m above the HAP. The BBOR sites were also characterized, in general, by adiabatic bottom layers although the lack of strong negative temperature gradients in the bottom several hundred meters here precluded formation of a "lid" at the top of these layers, except possibly on STD casts 4 and 9. (Station 4 was misplaced considerably downslope from the mean position of E, as seen by its greater depth (Figs. 3, 16). The bottom temperatures at station 4 and the profile shape indicate an AABW regime, more typical of the HAP stations.). Instead, relatively broad temperature minima at 150 to 300 m above bottom mark the upper limit of the adiabatic water layer. These adiabatic bottom layers on the BBOR and the HAP are both interpreted as zones of thorough mixing with respect to temperature despite the difference in the sharpness of the temperature minimum defining the top of this layer. It is assumed that this thorough mixing is produced by turbulence in the bottom layer. An alternative interpretation of these adiabatic layers is that they are due to geothermal heating from below. This is regarded as unlikely in this boundary current regime, where relatively high current velocities supply the necessary shear to drive the turbulence.

The weak temperature gradients observed on the BBOR stations imply weak density stratification, since no strong salinity gradients are present here. Unfortunately, stability cannot be computed quantitatively without more precise salinity data.

Nephelometer and Suspended Particulate Profiles

A composite of all the nephelometer profiles obtained at stations E, F, and G is given in Figure 3. These are typical of most deep-ocean profiles in that they show a steady decrease from the surface downward then an increase as the bottom is approached, forming a light-scattering minimum at around 3400 m at E and F and around 4500 m at G. On E and F, from 2500 to 3500 m, the

light scattering was constant for the 19-day period. The range of
values encountered for a given depth can be taken as an estimate of
the "worst" measurement precision. This value, ±.05 in log S,
where $S = E/E_D$, is a "worst" estimate because the variations may
indeed be real changes in the water column.

Below the scattering minimum at the BBOR sites — that is, in
the nepheloid layer — high scattering values were recorded and
large variations were observed between casts with near-bottom log-S
values ranging from 1.0 to 1.8. In contrast, the scattering profiles in the nepheloid layer at the HAP sites were quite constant,
with moderate near-bottom light-scattering values ranging from 0.9
to 1.2. The variations observed at the BBOR sites extend up to
1.5 km above the bottom. Although the scattering recorded in the
upper part of the nepheloid layer (3500-4000 m depth) represents
very small concentrations of suspended material (on the order of
10 µg/ℓ or less), they are real deviations from "background" turbidity or scattering, that is, they are significantly above the detection level of the instrument. Thus the height of approximately
1.5 km above bottom apparently represents the upper limit at which
the water-column particulate population is affected by the frictional interaction of the bottom-water mass with the ocean bottom.

High gradients in the profiles are observed at a height of
about 150 to 300 m above bottom at the BBOR sites and at about 80 m
at the HAP site. Keeping in mind the logarithmic scattering units
in Figure 3, it can be appreciated that a large majority of the
scattered flux recorded is from the bottom layer below this high
gradient. This point is emphasized in Figure 4 where the scattering
values in the bottom 1800 m of water are shown on a linear scale.
The height of the temperature minimum from Figure 2 has been noted
on the corresponding nephelometer profiles in Figure 4. This level
marks the top of the adiabatic (well-mixed) layer above the BBOR,
corresponds to the "lid" on the 60 to 80-m thick adiabatic bottom
layer above the HAP and corresponds to the level of maximum turbidity gradient. The significance of this correspondence will be discussed below.

The same correlation between the height of the temperature minimum and the maximum gradient in particulate concentration can be
seen in the profiles of suspended particulate concentrations (Figures 7, 8 and 9) although less clearly than on nephelometer profiles
because of sample spacing.

The gross differences between the level of suspended particulate concentrations and their vertical distributions are shown in
Figure 10, in which all values for each station have been plotted.
Note the higher concentrations and the greater variability of the
distributions at stations E and F compared to station G.

The variability with time in near-bottom suspended particulate
concentrations is also seen in Figures 7, 8 and 9 in which measured
concentrations for discrete water samples are shown for each cast
at each station. Note that at stations E and F significant variability exists in the particulate concentration on the time scale
of successive casts (hours to days) not only immediately above the

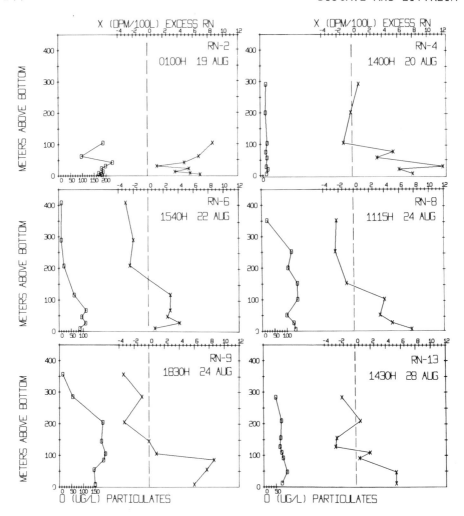

Fig. 7. Excess-radon/suspended particulate profiles at station E
on the Blake-Bahama Outer Ridge (cast Rn1 is not shown).

bottom but at heights of hundreds of meters above the bottom. This
corresponds to the nephelometer data. Although significant variability in light scattering extends up to 1.5 km above the bottom,
the significant variability in suspended sediment burden (by mass)
extends upward only several hundred meters. By contrast, both the
intensity, and the vertical dimension and degree of variability in
the suspended particulate concentrations at station G on the HAP
were much lower than that at stations E and F on the BBOR.

Integration of particulate concentrations over the near-bottom
water column yields a mass of suspended sediment per unit area, or
a particulate "standing crop." This parameter has been calculated
for each of the casts shown in Figures 7, 8 and 9, using 400 m above
bottom as the upper limit of the integration. This level is below

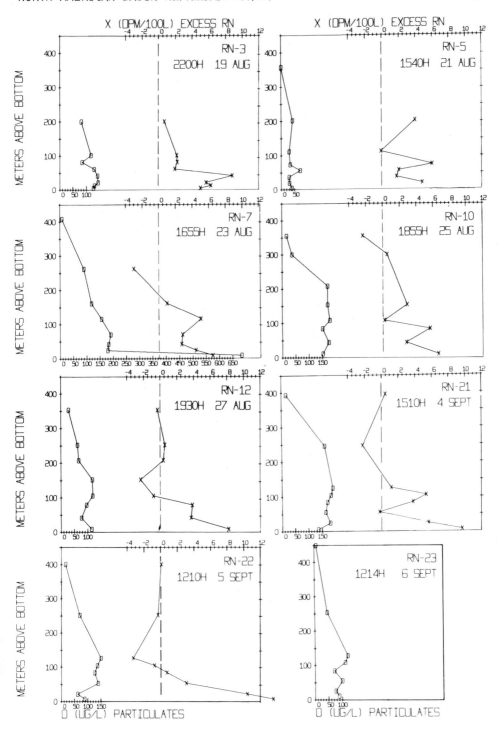

Fig. 8. Excess-radon/suspended particulate profiles at station F on the Blake-Bahama Outer Ridge.

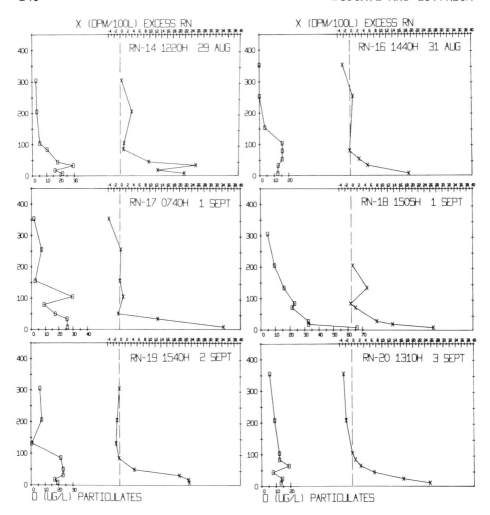

Fig. 9. Excess-radon/suspended particulate profiles at station
G on the Hatteras Abyssal Plain. Depths are given in meters.

the top of the nepheloid layer as defined by the turbidity minimum
but, as noted earlier, is above the level which includes the great
preponderance of suspended particulate matter by mass. Variations
in this parameter as a function of time are discussed below.

Excess-Radon Profiles

Comparison of the radon profiles at each station and among
stations reveals several consistencies (Figures 7, 8, 9 and 11).
First, both BBOR stations are characterized by a great variability
in the distribution of excess radon above the bottom while that at
the HAP station is significantly less variable. Variability includes the shape of the profiles (few at stations E or F approach

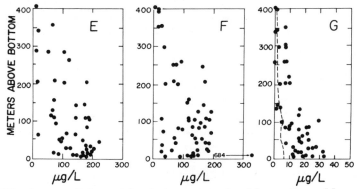

Fig. 10. Suspended particulate concentrations for all profiles from stations E, F and G. Dashed line = the upper limit of concentrations at station G were they plotted on the same horizontal scale used for stations E and F.

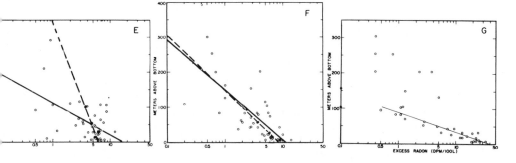

Fig. 11. Log of excess-radon versus height above bottom for profiles at stations E, F and G. Solid line = the regression curve on all points; dashed line = the regression curve on the mean concentrations for stratified intervals described in the text; solid points = the mean concentrations for the several intervals.

the simple exponential decrease; most at station G do), and the concentration of excess radon at the bottom (concentrations are lower and more variable at stations E and F than at station G). The vertical distributions at stations E and F suggest a more vigorous dispersal of radon higher into the water column than at station G. Whereas, only at station G do individual profiles approach an exponential decrease above bottom, taken together; that is, integrating over time, there is also some suggestion of an exponential decrease with height at station E and more so at station F. A distribution which fits such an exponential model would, on semi-log paper, plot along a straight line with a slope proportional to vertical mixing or eddy diffusivity. The three collective distributions are plotted in Figure 11. These semi-log plots confirm two conclusions. First, that stations E and F on the BBOR display considerable variability which, integrated over ten and eighteen days, respectively, are not convincingly exponential in shape. Assuming,

however, that the radon distributions are due to diffusion from the
bottom, and forcing this model; that is, attempting to "fit" a line
to the points for stations E and F in Figure 11, results in eddy
diffusivities of 33 cm^2/sec for station E (with a correlation coefficient of r = -0.52) and 102 cm^2/sec for station F (with r =
-0.72). Professor Claes Rooth (personal communication) has pointed
out, however, that a more legitimate treatment of the data would be
stratification in height intervals, computing the average concentrations for each interval, and fitting the eddy diffusion model to the
height-averaged data. The rationale for this is that, if the time
intervals between samplings for radon are longer than the characteristic eddy time scale, effectively independent realizations of a
random process are being observed, and such a situation should be
analyzed in terms of ensemble averages. Such a treatment also has
the effect of minimizing sampling bias at certain heights above the
bottom. If the variations within the strata are small, the difference between the two treatments is not important; but this is not
the case, particularly at station E. Computing average concentrations for both stations E and F within the height intervals of 0-10,
10-20, 20-40, 40-80, 80-120 and 120-160 m above bottom, these
averages and the regressions on them have also been plotted (dashed
line) in Figure 11. Eddy diffusivities calculated on the stratified
data are 680 cm^2/sec (r = -0.95) at station E and 260 cm^2/sec
(r = -0.98) at station F. The diffusivities by both treatments are
more comparable at station F 100 and 260 cm^2/sec) but much higher at
station E by the stratified-data treatment (33 vs. 680 cm^2/sec). By
either treatment, however, the vertical mixing indicated by radon at
stations E and F on the BBOR is significantly greater than at station G on the HAP. All the data at station G on the HAP in the lower 80-100 m (whether taken as individual data points or stratified)
fit an exponential model with a calculated eddy diffusivity of
about 12 cm^2/sec. As seen from Figure 11, these data fit an exponential model much better without stratifying the data (r = -0.94)
than for stations E or F. (Excess radon above 100 m was not included in the calculation of eddy diffusivity for station G.) Whether
or not the exponential model is rigorously applicable, the excess-
radon data require significantly more vertical mixing by some
mechanism at stations E and F than at G.

The excess-radon profiles, especially when integrated over
time, show direct relationships between degree of vertical mixing
and the quantity of suspended sediment in the water column.

Bottom Current Measurements

The progressive vector diagram in Figure 12 shows the theoretical path a water particle would take starting from station E on 18
August and progressing to 28 August, based on the current magnitudes
and directions recorded at station E by the Geodyne current meter.
The direction is remarkably constant at 150° from magnetic north

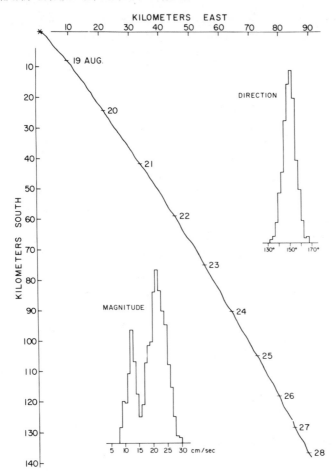

Fig. 12. Progressive vector diagram derived from the 10.5-day current measurement at station E, one meter above the bottom. Coordinates are relative to magnetic north, 7° west of true north.

(with tidal fluctuations superimposed) with predominant magnitudes at 23 cm/sec, ranging up to 30 cm/sec. The direction of the bathymetric contours is about 167° from true north, or 174° from magnetic north (Figure 1), approximately parallel to the current direction recorded during the last few days of the recording period. Thus, the current, interpreted to be the WBUC, is predominantly along the contours but with a significant downslope component at the beginning of the recording period. This downslope component of direction is consistent with the direction of an Ekman spiral, were one present in the benthic boundary layer. The period of downslope flow is also the period of highest current magnitude. Clearly, if the smooth variation observed during the 10.5-day recording is part of a periodic phenomenon, a longer recording time is necessary to specify it. The histograms in Figure 12 show the relative constancy of current direction and suggest a bimodality in current velocity.

The important point to note here is the relative constancy of
the motion with no sudden accelerations in direction or magnitude;
that is, the variability in suspended particulate and excess-radon
profiles is not a simple function of current velocity.

Bottom Photographs

The differences between the BBOR and HAP stations in intensity
and variability of vertical mixing is matched by evidence from the
bottom photographs. At the BBOR sites, the bottom was characterized
by prominent scour and reworking of the sediments. Scour-produced
ledges, scour "moats" around rocks, "crag and tail" features,
textured lineation of the bottom and ripples were found in the
vast majority of the photographs (Figure 13): *Heezen and Hollister*
[1972] give additional examples of some of these features and their
relation to bottom currents. Current directions indicated from
these features ranged from 140°-160° (magnetic) with a mean of
150° at station E and 133°-157° with a mean of 147° at station F.
Also, sedimentary rock outcrops were photographed at 3 of the 6
casts at station E. The relief associated with the rock outcrops,
as an indication of bed roughness, ranged up to about 1/2 m. In
contrast, all the photographs at the HAP site displayed a monotonous,
flat, mud bottom showing some lebenspuren with a relief of about
1 cm. Very slight destruction by current action on the lebenspuren
was seen in some photographs but no current lineations or scour
were visible. In two photographs from the HAP, some indication of
northward current flow can be derived but that conclusion is somewhat tenuous. In the first photograph, the mud cloud stirred up
by the trigger weight was seen to stream slightly northward.
In the second, one can distinguish the leeward and windward sides of
a lebenspuren in the process of being destroyed by current erosion,
also indicating northward flow. Thus, the only information on
current direction on the HAP, though limited, indicated weak
northward flow.

Summary of Characteristics of Stations E, F, and G

In preceding sections the results of several types of measurements at stations E, F, and G have been discussed separately. The
behavior at the three stations of almost all of the parameters are
interrelated and can be summarized as follows.
Stations E and F on the BBOR are characterized by high
scattering and suspended-particulate concentration values and
a high degree of variability in their vertical distributions.
These variations extend high into the water column, the great
bulk by mass varying below 300 m but some variability extending up
to ∿ 1500 m above bottom (∿ 3500 m water depth). Particulate
standing crops (concentrations integrated up to 400 m above bottom)
also display large variability. The maximum height to which the

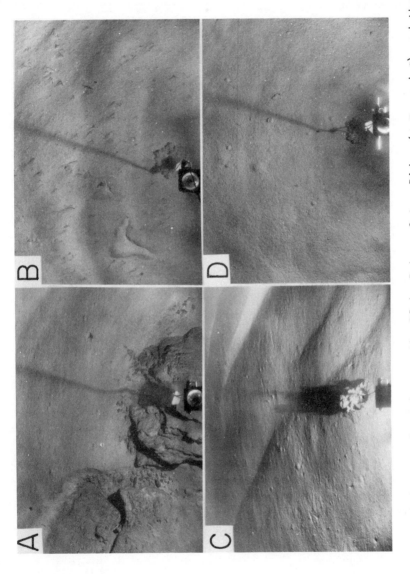

Fig. 13. Bottom photographs taken on the Blake-Bahama Outer Ridge (*A*,*B*, and *C*) and the Hatteras Abyssal Plain (*D*). The width of the field of view is approximately 2 m. In *A* and *B* current direction can be seen relative to the compass needle, pointing to 4 o'clock and 6 o'clock, respectively.

particulate matter is carried, i.e. ∿ 1500 m (minimum scattering
on nephelometer profiles), corresponds to a zone of inflection
points in the temperature-versus-depth profiles below which the
negative temperature gradient decreases. (These inflection points
cannot be seen in Figure 2 in which only the lower portion of the
STD analog records are included.) The second derivative of temperature, with respect to depth, is positive from these inflection points
downward; that is, the profile is concave to the right. We infer
that these inflection points mark the deepest level of which
typical midwater conditions of low diffusion rates prevail.
Below this level, the effect of bottom friction becomes apparent
and mixing rates increase, giving rise to increasingly homogenized
water with depth. In general, a temperature minimum exists from
150-300 m above bottom, below which temperature increases adiabati-
cally. This temperature minimum also marks the zone of high gradient
in light scattering below which most of the suspended particulate
burden exists (Figure 4).

Temperature (and salinity) characteristics of this water mass
are consistent with its identification as water of high-latitude
North Atlantic sources and the 10.5-day current meter record at
1 m above the bottom at station E is consistent in direction
(∿155°) and velocity (20-30 cm/sec) with the interpretation that
this current is the WBUC. From photographs, the nature of the
bottom shows local relief or roughness on the order of 1/2 m and
shows significant scouring of the bottom. Vigorous near-bottom
turbulence is also evidenced by the distribution of excess radon
which is homogeneously mixed rather high into the water column
(Figures 7. 8, 11). The excess radon shows a great variability in
vertical distribution including relatively low near-bottom concentrations and the presence of above-bottom maxima.

Thus, all parameters measured at stations E and F indicate a
rapidly moving, turbulent water mass which is scouring the bottom
to support the nepheloid layer, a zone of relatively turbid water in
which the bulk of the mass or standing crop lies within 300 to 400
m of the bottom but which is manifest as high as 1500 m above the
bottom. The other salient characteristic of the stations on the
BBOR is the short term variability of those parameters which reflect
interaction of the turbulent water with the bottom. These variations
are discussed in the section following this summary.

In contrast to the BBOR, station G on the HAP was characterized
by much lower values of light scattering and suspended particulate
concentrations, although their increase very close to the bottom
must be termed a nepheloid layer, albeit a relatively weak one (Figures 3, 4, 9, 10). The vertical distribution of suspended particulates was much less variable than on the BBOR and the level of
clearest water above the HAP (representing the highest influence
of bottom sediment sources, that is, the top of the nepheloid layer)
was only ∿ 900 m above bottom at ∿ 4500 m depth (cf. 1500 m above
bottom and 3500 m depth at station E and F) (Figures 3, 4 and 10).
This depth of 4500 m also corresponds to a zone of inflection
points in the profiles of temperature versus depth from a positive
to a negative second derivative, that is, concave to right above,

concave to left below (Figure 2). This inflection point marks
the upper limit of immediate diffusive influence of the cold bottom water flowing in from the south in the HAP. This water mass
adjacent to the bottom at station G is the AABW ($\sim$ 20% according
to Wüst's criterion (*Wüst* [1933]; *Amos et al.* [1971])) and,
although current measurements are lacking, bottom photographs
indicate tranquil conditions at the sediment/water.interface
(Figure 13). This relative tranquillity corroborates indications
of vertical eddy diffusivities from excess-radon profiles of
$\sim$ 10 cm^2/sec within the bottom 80-100 m. This 80-m zone also
approximates that of the near-bottom adiabatic layer seen in all
three STD profiles on the HAP (Figure 2). The high gradient "lid"
on this adiabatic layer is also coincident with a high gradient
in suspended particulates (Figure 4).

Thus all parameters at station G suggest a more tranquil, less
turbulent water mass adjacent to the bottom, resulting in a much
less intense nepheloid layer and a less variable vertical distribution of bottom-source parameters with time.

Temporal Variability

Of all the measurements made repeatedly at each station, no
two casts displayed identical results. As summarized in the
previous section, this variability was much lower (perhaps insignificant) for all parameters at station G than at stations E and
F, as seen in Figure 14 in which are plotted, as a function of
time, several parameters related to the intensity of the nepheloid
layer: turbidity and suspended particulate concentration 10 m
above bottom; integrated suspended particulate matter — that is,
"standing crop" in mg/cm^2; potential temperature; and, at station
E, current velocity and direction 1 m above bottom. From this plot
several conclusions concerning the observed variability can be
drawn. (*1*) Variations in the intensity of the nepheloid layer
during the experiment were real and outside the limits of experimental error. (*2*) The variability was not random, but apparently
was organized in a regular manner with time. From the every-
other-daily rate at which each station was sampled, it is possible
that large variations occurred much more frequently than the
curves shown in Figure 14; however, this possibility is regarded
as highly unlikely. (*3*) The observed variations were approximately
in phase ($\pm$ one day) at stations E and F, which were separated by
110 km distance. (*4*) If there is a real cyclicity in the variations,
it appears to have a period of about 1 week. Note that this does
not correspond to any obvious period in current velocity or direction. (*5*) The observed variations at stations E and F were not
a function of slight changes in location; that is, because of
surface drift, it was not possible to position the ship and
sampling package in exactly the same location on successive
reoccupations of each station. This is shown in Figure 15, in
which particulate standing crop is plotted versus water depth, as
an indication of geographic variability; that is, if the core of the

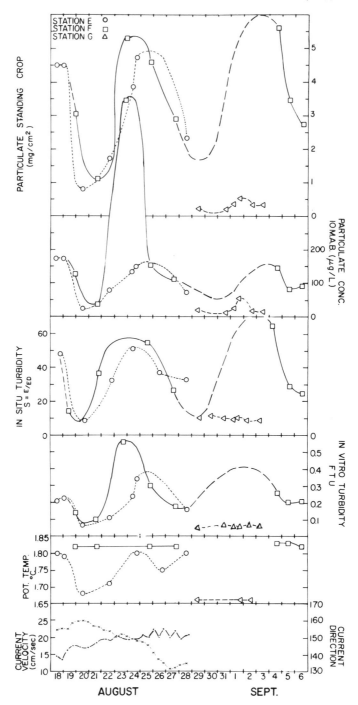

Fig. 14. Temporal variations in parameters observed in the benthic boundary layer. O = E values; □ = F values, with long dashes indicating variation in casts on 27 Aug and 4 Sept; Δ = G values.

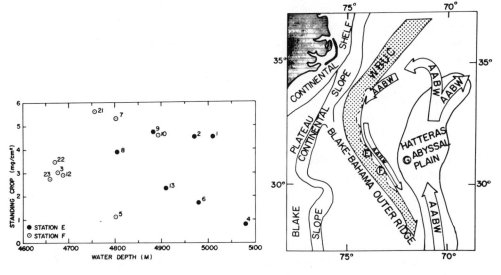

Fig. 15. Particulate standing crop (mg/cm^2) vs. water depth (m) for stations E and F.

Fig. 16. Schematic representation of the interaction of relatively clean Antarctic Bottom Water (AABW) with the Western Boundary Undercurrent (WBUC) to form the variations in turbidity observed at stations E and F.

WBUC were fixed in location along the slope of the BBOR and the observed variations represented merely the proximity of sampling to that core, a linear or curved correlation would be shown between the two parameters for either station singly or together. From the random scatter of the data points in Figure 15, it is concluded that such was not the case and the degree of geographic variation experienced did not affect results. Figure 15 does, however, show that the average depth at station F was less than that at station E.

A dynamic model is presented (Figure 16) to explain the temporal variability observed on the BBOR. This model, although recognized as imperfect, is compatible with the data. Pulses of relatively low-turbidity water are envisioned as being injected into the more turbid WBUC upstream from sampled stations and being carried downstream at boundary current velocities such as those observed at station E. As a mechanism for generation of the low turbidity pulses an interfingering of southeastward flowing AABW with the main WBUC at the northern end of the HAP is postulated.

The AABW regime on the HAP is characterized by lower turbidity and by about 0.15° C-lower potential temperature than the WBUC regime on the BBOR. Because of a suggestion of correlation between turbidity and water temperature at station E, the deeper of the two stations on the BBOR (warmer temperature with higher turbidity, Figure 14) it is attractive to attribute the observed fluctuations to water mass differences. The temperature/salinity sections of

Amos et al. [1971] show two water masses comprising the southeast-ward flow on the BBOR flank. The main flow at about 4500 m is the Lower North Atlantic Deep Water (NADW), but farther down the slope, centered on about 5300 m, is a core of recirculated (south-ward flowing) AABW. *Amos et al.* [1971] propose that this AABW has circulated counterclockwise around the northern end of the HAP, being restricted from flowing further north by the topography. The boundary between these two water masses is at about 5000 m, close to the depth of our stations on the BBOR. we suggest that, at the northern end of the HAP, boluses of AABW are injected into the boundary region between the WBUC and the AABW or, at least, distort this boundary region with a series of eddies frictionally coupling the two water masses. Thus, at stations E and F, located near the depth of this boundary region, alternately warm-turbid and cooler-clearer water were observed as these eddies were transported southeastward by the WBUC.

In order to account for the in-phase nature of the variations of stations E and F, however, the distance between these pulses or eddies must be approximately the same as the distance between E and F; that is, the pulses must be generated with approximately a one-week period. This is the order of both the travel time between stations E and F ($\sim$ 6 days for the observed mean velocity of $\sim$ 22 cm/sec) and the observed apparent period in our data.

The fortuitous aspect of the dynamic model presented is that the frequency of pulses of turbidity should happen to be such that the leading edge of a pulse would arrive at station E simultaneously with the arrival at station F of the leading edge of the previous pulse. Another problem with the model is that, although the variations in turbidity at station F are at least as great as those at station E, the bottom temperature values at station F (although not sampled as well in time as at E) do not vary sympathetically with turbidity as at station E.

This model implies a non-local origin of the suspended particulate matter in the WBUC in that the less turbid eddies of AABW-like water retained their relative clarity over several hundreds of kilometers.

CONCLUSIONS

As a result of almost three weeks of observations of phenomena in two regions of the benthic boundary layer of the western North American Basin, the following conclusions are drawn:

1) The character of the benthic boundary layer was distinctly and reproducibly different in the two regimes studied. In the regime of the Hatteras Abyssal Plain, the layer was thinner and less variable than in the Western Boundary Undercurrent regime of the Blake-Bahama Outer Ridge.

2) The several characteristics of the benthic boundary layer observed in each regime were interrelated. The near-bottom water temperatures and salinities at stations E and F on the BBOR as well as the current velocities and directions recorded at station E

were consistent with the water mass there being Lower North Atlantic Deep Water. Consistent with this high-velocity regime, the
suspended particulate matter was high and quite variable in concentration, as measured by *in situ* nephelometry, *in vitro* turbidity
and by gravimetric analysis of filtered samples. The vertical
distributions of excess radon were also consistent with a high-
velocity regime in that near-bottom concentrations were generally
low and vertical distributions were extremely variable. The
variations were such that low and high concentrations were
equally likely to occur at 10 or 100 m above the bottom, suggesting a fairly random mixing process with eddy sizes sufficiently
large that, within times of the order of magnitude of the radon-
decay time (4 days), little degree of homogenization is achieved.
The vertical eddy diffusivity calculated on the data at station
F equals 100 cm^2/sec. The vertical distributions of particulate
matter, radon, and temperature show interrelationships in the
coincidence of the high gradient in suspended particulate matter
with a temperature minimum between 150 to 300 m above bottom.
The zone below this temperature minimum also includes almost all
the excess radon. In addition, the top of the nepheloid layer
(scattering minimum) is coincident with an inflection point in
the temperature gradient at around 3500 m water depth or 1500 m
above bottom, which may mark the upper limit of thermal mixing of
bottom water. Evidence of high water velocity in the direction
of the WBUC, high turbulence, and resulting erosion is seen in
the bottom photographs which, at stations E and F, show scour
marks, current ripple marks, rock outcrops and other indications
of bottom roughness consistent with the other observations.

By contrast, the measurements at station G on the HAP showed
low concentrations of suspended particualte matter and little
variability in their vertical distributions. Excess-radon
profiles were also less variable and yielded an over-all eddy
diffusivity an order of magnitude lower than that calculated for
station F on the BBOR. As on the BBOR, the high gradient in
both radon and suspended particulate concentrations about 80 m
above bottom is coincident with a sharp break and a maximum in
the temperature gradient and, assuming small changes in salinity,
probably represent a strong maximum in water-mass stability.
The top of the nepheloid layer (scattering minimum) was at 4500 m
water depth or about 900 m above bottom and is coincident with
the upper limit of thermal mixing of AABW as denoted by an inflection point in the temperature-versus-depth gradient. These
interrelated phenomena are consistent with a much lower belocity,
lower turbulence regime in which a bottom benthic layer is formed
in Antarctic Bottom Water which is presumably moving north over
the HAP. The lower velocity and turbulence are deduced from
bottom photographs which show relatively undisturbed lebenspuren
and none of the erosion features common to the BBOR.

3) The phenomena measured in the benthic boundary layer of the
HAP at station G showed little or no variability during the weeks
observation. The variability at E and F on the BBOR, however, was
large and apparently regular. The intensity of the nepheloid layer

10-20 m above bottom as measured by *in situ* and *in vitro* turbidity, and by gravimetric analysis of filtered particulate matter, varied over significant factors with an apparent periodicity of about one week. The standing crop of particulate matter, less subject to the extreme variations of measurement of a single sample, also varied by about a factor of five within the same period. These parameters varied with the same apparent period and were in phase (± one day) at stations E and F. The data at station E also indicated a positive correlation between near-bottom potential temperature and suspended particulate concentration, a correlation which suggests that the variability was related to water-mass characteristics. No such variation in near-bottom potential temperature was observed at station F, however, despite the comparable variability in nepheloid-layer intensity and simultaneity with fluctuation at station E.

   4) A tentative explanation of the variability observed on the flank of the BBOR is suggested. A portion of low-turbidity, cold AABW is deflected southwestward by the topography at the northern end of the HAP. Where this current impinges on the southward flowing WBUC, boluses of AABW are incorporated into the WBUC and carried south along the flank of the BBOR. These were observed at stations E and F as fluctuations in the intensity of the nepheloid layer.

ACKNOWLEDGMENTS

   Gratitude is expressed to Captain Allen L. Jorgensen and the officers and crew of the R/V *Robert D. Conrad* for their excellent cooperation during this experiment. We thank John Harvey, Roy Wilkins and Anthony Amos for the STD and current-meter data; their efforts and equipment were supported under Atomic Energy Commission contract AT(11-1)2185. We thank Adele Hanley, Wayne Bottner and Larry Carroll, Jr. for technical assistance. For critical reading of the manuscript and for helpful suggestions we are grateful to Wallace Broecker (Lamont-Doherty Geological Observatory), Abe Lerman (Northwestern University), Wolfgang Roether (Heidelberg University and Lamont-Doherty Geological Observatory) and Claes Rooth (University of Miami). We gratefully acknowledge funding of this project by the Office of Naval Research under contract N00014-67-A-0108-0004, by the National Science Foundation under grant GA27281 and by the Atomic Energy Commission under contract AT(11-1)3132.

   This paper is contribution 2052 of the Lamont-Doherty Geological Observatory.

REFERENCES

   Amos, A., A. Gordon, and E. D. Schneider, Water masses and circulation patterns in the region of the Blake-Bahama Outer Ridge, *Deep Sea Res., 18,* 145-165, 1971.

Broecker, W. S., An application of natural radon to problems in ocean circulation, in *Symposium on Diffusion in Oceans and Fresh Waters*, edited by T. Ichiye, pp. 116-144, 1965.

Broecker, W. S., J. Cromwell, and Y. H. Li, Rates of vertical eddy diffusion near the ocean floor based on measurements of the distribution of excess 222_{Rn}, *Earth and Planet. Sci. Letters*, vol. 5, pp. 101-105, 1968.

Broecker, W. S., Y. H. Li, and J. Cromwell, Radium-226 and Radon-222: Concentration in Atlantic and Pacific Oceans, *Science, 158*, 1307-1310, 1967.

Chung, Y., and H. Craig, Excess radon and temperature profiles from the Eastern Equatorial Pacific, *Earth and Planetary Sci. Letters*, vol. 14, pp. 55-64, 1972.

Connary, S. C., and M. Ewing, The nepheloid layer and bottom circulation in the Guinea and Angola Basins, in *Studies in Physical Oceanography*, edited by A. L. Gordon, Gordon and Breach, London, pp. 169-184, 1972.

Eittreim, S., P. Bruchhausen, and M. Ewing, Vertical distribution of turbidity in the South Indian and South Australian Basins, in *Antarctic Oceanology II: The Australian-New Zealand Sector*, edited by D. E. Hayes, Amer. Geophys. U., pp. 51-58, 1972a.

Eittreim, S., M. Ewing, and E. M. Thorndike, Suspended matter along the continental margin of the North American Basin, *Deep Sea Res., 16*, 613-624, 1969.

Eittreim, S., and M. Ewing, Suspended particulate matter in the deep waters of the North American Basin, in *Studies in Physical Oceanography*, edited by A. L. Gordon, Gordon and Breach, London, pp. 123-167, 1972.

Eittreim, S., A. L. Gordon, M. Ewing, E. M. Thorndike, and P. Bruchhausen, The nepheloid layer and observed bottom currents in the Indian-Pacific Antarctic Sea, in *Studies in Physical Oceanography*, edited by A. L. Gordon, Gordon and Breach, London, pp. 19-35, 1972b.

Ewing, M., and S. Connary, Nepheloid layer in the North Pacific, in *Geol. Soc. Am. Memoir 126*, edited by J. D. Hays, pp. 41-82, 1970.

Ewing, M., S. Eittreim, J. Ewing, and X. LePichon, Sediment transport and distribution in the Argentine Basin: Part 3, Nepheloid layer and processes of sedimentation, in *Physics and Chemistry of the Earth*, vol. 8, pp. 49-77, 1971.

Ewing, M., and E. M. Thorndike, Suspended matter in deep ocean water, *Science, 147*, 1291-1294, 1965.

Fuglister, F. E., *Atlantic Ocean Atlas of temperature and salinity profiles and data from the International Geophysical Year 1957-58*, Atlas Ser. Woods Hole Oceanographic Inst., vol. 1., 1960.

Heezen, B. C., and C. D. Hollister, *The Face of the Deep*, Oxford University Press, New York, 659 pp., 1971.

Heezen, B.C., and M. Tharp, Physiographic diagram of the North Atlantic Ocean (revised 1968 *for* The Floors of the Oceans), *Geol. Soc. Amer. Spec. Paper 65*, 1968.

Hunkins, K., E. M. Thorndike, and G. Mathieu, Nepheloid layers and bottom currents in the Arctic Ocean, *J. Geophys. Res., 74*, 6995-7008, 1969.

Lee, A., and D. Ellet, On the water masses of the Northwest Atlantic Ocean, *Deep Sea Res.*, *14*, 183-190, 1967.

Sternberg, R. W., Field measurements of the hydrodynamic roughness of the deep-sea boundary, *Deep Sea Res.*, *17*, 413-420, 1970.

Swallow, J. C., and L. V. Worthington, An observation of a deep counter-current in the western North Atlantic, *Deep Sea Res.*, *8*, 1-9, 1961.

Thorndike, E. M., A deep sea photographic nephelometer; in preparation, 1974.

Thorndike, E. M., and M. Ewing, Photographic nephelometers for the deep sea, in *Deep-Sea Photography*, edited by J. B. Hersey, Johns Hopkins Press, Baltimore, pp. 113-116, 1967.

Wimbush, M., and W. Munk, The benthic boundary layer, in *The Sea*, vol. 4, part 1, edited by A. Maxwell, Wiley Interscience, New York, 731-758, 1970.

Wüst, G., Das Bodenwasser und die Gliederung der Atlantischen Tiefsee, *Wiss. Ergebn. dt. atlant. Exped. Meteor 1925-27*, vol. 6 (1) (1), 106 pp., 1933. (English translation No. 340 by M. Slessers, U.S. Naval Oceanographic Office, 1967)

Distribution of Suspended Particles in the Equatorial Pacific Ocean

HASONG PAK

Oregon State University

ABSTRACT

Light scattering, size distribution of the suspended particles and standard hydrographic parameters were measured on water samples collected at 152 stations in the Panama Basin during October and November 1971. The particle-volume concentrations are calculated by fitting the measured particle size distributions to the exponential distribution and integrating the exponential distribution. Total suspended matter was calculated using a particle density of 2.2 mg/ml. The suspended matter in the water below 25 m depth ranges from 0.04 mg/l to 0.092 mg/l and from 0.11 mg/l to 1.76 mg/l in the sea surface.

The spatial distribution of the suspended matter appears to be closely related to the dynamic conditions of the water: the equatorial front, interaction of the Cromwell Current and the Galapagos Islands and interaction of bottom current against the Carnegie Ridge.

INTRODUCTION

Sea water contains a large number of solid particles. Their importance to many oceanographic problems is widely recognized but observed data are still lacking. The few observed data indicate that the particle-size distributions are nearly exponential [*Lisitzin*, 1961; *Ochakovsky*, 1966; and *Plank et al.*, 1972]; the larger particles contribute most of the volume although the greatest number of particles seem to be smaller than 1 μ. The suspended particles in open and deep ocean water range, by weight, from about 1 μg/l to 200 μg/l [*Jacobs and Ewing*, 1969; *Harris*, 1972].

Particle-size analysis is a difficult and tedious operation, often impossible at sea. Data obtained by different investigators are not readily comparable because different methods are often used and rarely are the sampling points common. Furthermore, there are uncertainties about how long a sample of water can be preserved without affecting the natural size distributions of the particles. *Parsons* [1963] has discussed several methods of determining the fraction of the particulate organic matter in sea water.

Particle-size distributions are determined traditionally by microscopic counting. *Lisitzin* [1961], *Ochakovsky* [1966] and *Harris* [1972] found a predominant number of particles in the small size ranges using the microscopic method in counting the distributions of the particles in the oceans. *Bader* [1970], *Beardsley et al.* [1970] and *Carder et al.* [1971] used the Coulter counter to determine particle-size distributions. The use of this counter has been discussed extensively by both *Sheldon and Parsons* [1967] and *Carder* [1970]. The reported particle-size distributions based on Coulter counter technique are the Junge distribution [*Bader*, 1970], the exponential distribution [*Carder et al.* 1971], and the log-log distribution [*Brun-Cottan and Ivanoff*, 1970].

The total mass of suspended matter is often determined by direct gravimetric measurements of samples collected by filtration or by centrifuge methods [*Ewing and Thorndike*, 1965; *Harris*, 1972; and *Lisitzin*, 1972]. The main difficulty with gravimetric measurements lies in the uncertainty or low accuracy due to the extremely low concentration of the suspended particles. Filtration of larger volumes of sample water is necessary to overcome this difficulty, but it makes the method more cumbersome and possibly more susceptible to contamination.

This study presents data concerning particle-size distribution and light scattering β_{45} measured in the Panama Basin using a Coulter counter and a Brice-Phoenix light-scattering photometer. Light-scattering data are a good indicator of total particle concentration if the suspended matter is fairly uniform in composition and size distribution. Such conditions may prevail in open oceans at depths greater than about 300 m. Light-scattering data are important in theoretical studies involving the inherent optical properties of sea water in addition to being a simple indicator of total particle concentration. Some of the relationships between light scattering and particle concentrations in the deep sea have been discussed by *Pak et al.* [1971] and *Beardsley et al.* [1970].

EXPERIMENTAL PROGRAM

Measurements were made at 152 stations in the Panama Basin during the period from 14 October to 7 December 1971 on Oregon State University's R/V *Yaquina*. Locations of the stations are given in Figure 1. Light-scattering measurements on samples obtained using NIO bottles were made using a Brice-Phoenix light-scattering photometer [*Spilhaus*, 1965; *Pak*, 1970]. Particle-size distributions were determined in the particle-diameter range from 2.2 μm to 10.2 μm using a Coulter counter having a 100 μ aperture [*Carder*, 1970].

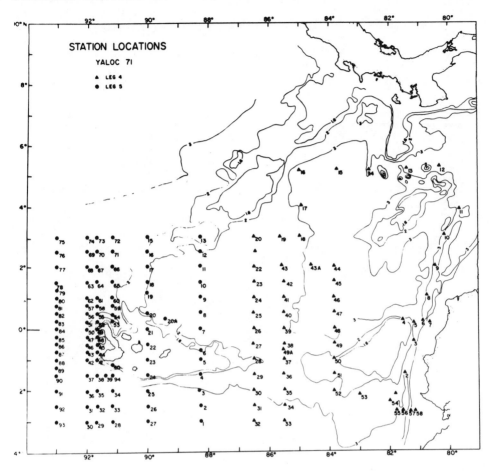

Fig. 1. R/V *Yaquina* Yaloc-1 cruise stations with bathymetry in km.

Light-scattering measurements were made in order to investigate the basic optical properties in addition to those specific applications presented here. β(θ) was measured at three different angles: 45°, 90° and 135° and with three different wavelengths: 436 nm, 546 nm, and 577 nm.

The volume concentrations of suspended particles were determined by fitting the observed points in the particle-size distribution to the exponential distribution and integrating the exponential distribution from 0 to infinity. There are indications that the shape of the size distribution of suspended matter in the ocean is relatively constant [*Bader*, 1970; *Plank et al.* 1972; *Sheldon et al.* 1972]. On the basis of the constant shape of the size distribution, the volume concentration determined by the Coulter-counter data and subsequent integration are believed to be reasonably accurate. The partial volume of the particles in the 2.2 - 10.2-μm diameter range determined by integrating the exponential distribution is in the range of 33 to 40% of the total volume.

RESULTS

   Cross sections of light scattering β_{45} and particle-volume concentration along the equator from 90°00'W to the coastal trench
(about 80°30'W), shown in Figure 2, demonstrate broad features of

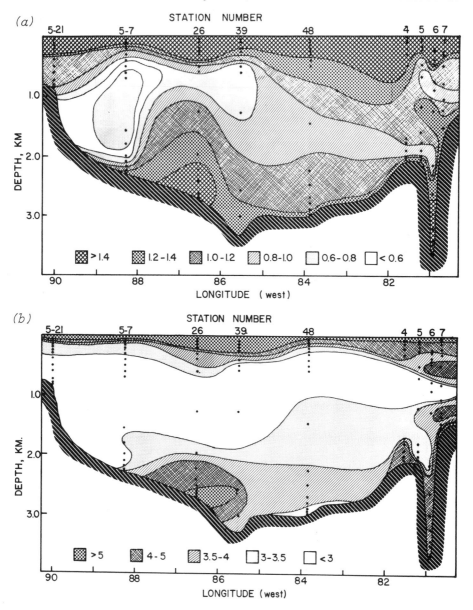

Fig. 2. Zonal cross sections of (a) the particle-volume concentration in ppm x 10^2 and (b) light scattering at 45° in $(m\text{-}str)^{-1}$ x 10^3
along the equator.

suspended particle distribution: a high concentration in the surface layer, a broad minimum at mid-depth and high values near the
bottom with maxima at both the coastal trench and at station 26
located near the saddle of the Carnegie Ridge.

Vertical distributions of light scattering and particle-volume
concentrations in the meridional section along 90°00'W for the upper
300 m of depth are shown in Figure 3. In the layer between 100 and
200 m, a relative maximum is found at about 01°00'S and 01°30'N,
which seems to be related to the Equatorial Undercurrent.

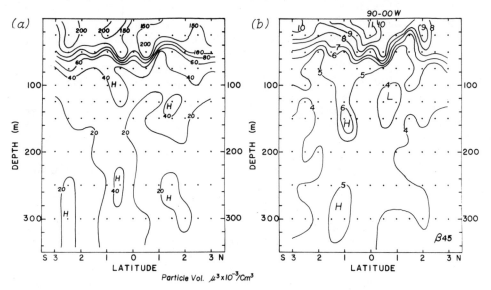

Fig. 3. Meridional cross sections of (a) the concentration of particles between 2.2 μm and 10.2 μm in ppm x 10^4 and (b) light scattering at 45° in (m-str)$^{-1}$ along 90°00'W.

The particle concentration data in Figures 2 and 3 can be used to
obtain an estimate of mass concentration. For an assumed particle
density of 2.2 mg/mℓ, 1 ppm by volume equals 2.2 mg/ℓ. The volume
concentrations of suspended particles range from about 0.018 to
0.042 ppm, corresponding to mass concentrations from 0.040 mg/ℓ to
0.092 mg/ℓ.

Particle concentration in the sea surface is shown in Figure 3.
A band of high particle concentration oriented almost zonally
through the Galapagos Islands and another branch extending toward
the southwest from the islands are evident. The total suspended
mass in the sea surface ranges from about 0.11 to 1.76 mg/ℓ.

DISCUSSION

Particle concentrations in terms of dry mass are more numerous
in the literature than any other parameter, probably because of the
demand for such information and the availability of the method of

determining concentration by filtration and weighing. *Jacobs and Ewing* [1969] reported particle concentrations from 0.05 x 10^{-2} mg/ℓ to 0.153 mg/ℓ with the average of 0.038 mg/ℓ for deep water in the North Pacific Ocean, and from 0.45 x 10^{-2} mg/ℓ to 0.087 mg/ℓ with the average of 0.03 mg/ℓ for deep water in the South Pacific Ocean. These values are approximately of the same order of magnitude as the average values of *Jacobs and Ewing* [1969].

The cross sections in Figure 2 show larger values in the volume concentration of suspended matter occurring at (1) the entire surface layer, (2) near the bottom in the coastal trench and (3) at station 26, located near the saddle of Carnegie Ridge. The surface layer of the ocean normally shows a high concentration of suspended matter due to (1) biological activities in the euphotic zone and (2) particle input from the atmosphere and from river drainage. The particle distribution in the surface of the ocean will be described latter.

The high concentration of particles in the coastal trench seems to be related to the proximity of the continent, which generally is considered a source of suspended particles. Erosion of coasts by waves, tides, currents, and drainage of river water are known sources of suspended particles near the continents. The high concentration of particles in the bottom water near the saddle at 85°W is interpreted as a result of bottom erosion by currents. This interpretation is supported by measurements of bottom sediment distribution and direct bottom current [*Van Andel et al.*, 1971].

Figure 3 is a detailed presentation of particle distribution in the upper 300 m. Two locations with relative maximum particle concentrations can be identified: one at about 1°S, extending from about 75 m depth and the other extending similarly at about 01°-30°N. Also identifiable is an area of relatively low values in particle concentration between the two maxima. Comparing Figure 3 to temperature distributions and distributions of oxygen concentration shown in Figure 4, we find a minimum vertical gradient in temperature and oxygen at the places where the maximum particle concentrations are indicated. The isolines of temperature and oxygen show sharp troughing below about 150 m depth and sharp ridging above 150 m depth. These types of temperature and oxygen distribution are known to be associated with the Equatorial Undercurrent [*Pak and Zaneveld*, 1973; *Montgomery*, 1962].

The maximum particle concentrations found in Figure 3 are interpreted as being associated with the Equatorial Undercurrent. The high concentration of suspended particles may be originated directly from the Galapagos Islands and carried downstream by the Equatorial Undercurrent. The water surrounding the Galapagos Islands has a high particle concentration (Figure 5), probably related to higher biological productivity in the area of the islands due, at least partly, to upwelling and strong internal mixing by the equatorial current system. A maximum particle concentration is also found in connection with the equatorial front [*Zaneveld et al.*, 1969] which is located approximately along the 0.2 ppm isoline near the equator (Figure 5).

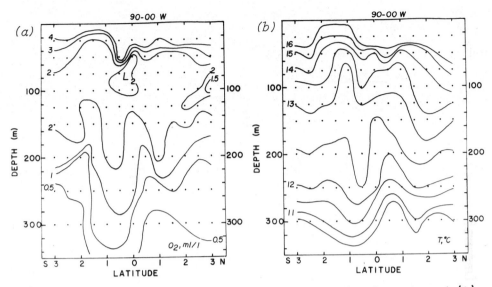

Fig. 4. Distributions of (*a*) oxygen concentration in mℓ/ℓ and (*b*) temperature (°C) along 90°00'W.

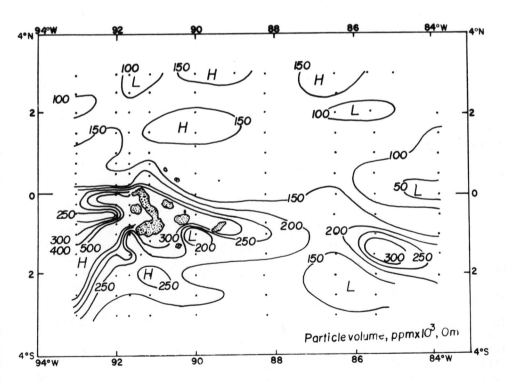

Fig. 5. The particle-volume concentration in the sea surface in ppm x 10³.

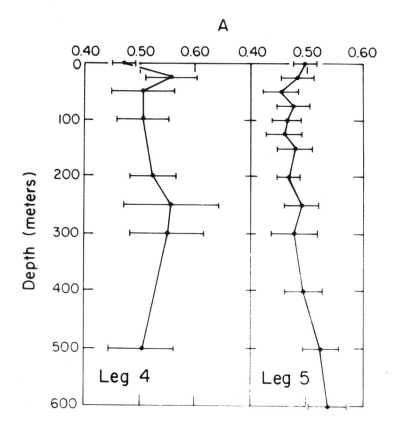

Fig. 6. Profiles of A which is a parameter of the exponential distribution, Ne^{-AX}.

With such a source of particles available in the area of the Galapagos Islands and along the equatorial front, the Equatorial Undercurrent, with its strong vertical mixing, is likely to cause the particle concentration seen in Figure 2.

Strong mixing due to the Equatorial Undercurrent is also found in the vertical distribution of a parameter A which is inversely related to a mean particle size [*Zaneveld and Pak*, 1973]. A is one of the two parameters of the exponential particle-size distribution, $f(D)dD = NAe^{-AD}dD$. In Figure 6, vertical distribution of mean A over 16 stations where the Equatorial Undercurrent is found (leg 5) is shown and is compared to similar data from stations not directly affected by the current (leg 4). Bars indicating 95% confidence level are also shown for the 16 stations and are compared to stations not affected by the current. It can be clearly seen that the lengths of bars at the Equatorial Undercurrent are much shorter than those at the other stations, indicating the size parameter A is much more uniformly distributed, probably due to the intense mixing in the Equatorial Undercurrent. There is a slight indication that A increases with depth at stations with the Cromwell Current (leg 5, Figure 6) while the other stations (leg 4, Figure 6) do not have the same indication. Such vertical distribution of A implies that mean particle size decreases with depth below the Cromwell Current and the mean particle size of the maxima is correlated with the Cromwell Current.

Through analysis of these data it was observed that the correlation between light-scattering β_{45} data and particle-size distribution varied considerably in space. Generally the correlation was higher where particle concentration was high. The significance of β_{45}, in terms of suspended particles, lies in the possible linear relationship between β_{45} and the total scattering coefficient b [Beardsley et al., 1970]. This relationship, although often assumed to be linear, may vary depending on composition and size distribution of particles. While light scattering is affected by all particles, measurements of particle size distribution in this study are limited to particle diameters between 2.2 and 10.2 μ. The constant shape of particle-size distribution may not be a valid assumption over the entire ocean; the shape may vary significantly from sources to sinks of particles.

ACKNOWLEDGEMENTS

The research for this study was supported by the Office of Naval Research through contract N00014-67-A-0369-0007.

REFERENCES

Bader, H., The hyperbolic distribution of particle sizes, *J. Geophys. Res.*, *75*, 2822, 1970.

Beardsley, G. F., Jr., H. Pak, K. L. Carder, and B. Lundgren, Light scattering and suspended particles in the eastern equatorial Pacific Ocean, *J. Geophys. Res.*, *75*, 2837, 1970.

Brun-Cottan, J. C., and A. Ivanoff, Particles' size distribution in sea water (abstract), presented at *"The Ocean World"*: *Joint Oceanographic Assembly*, IAPSO, IABO, CMG, SCOR, Tokyo, Japan, September 1970.

Carder, K. L., Particles in the eastern tropical Pacific Ocean: Their distribution and effect upon optical parameters, Ph.D. Thesis, Oregon State University, Corvallis, 1970.

Carder, K. L., G. F. Beardsley, Jr., and H. Pak, Particle size distributions in the Eastern Equatorial Pacific, *J. Geophys. Res.*, *76*, 5070, 1971.

Ewing, M. and M. Thorndike, Suspended matter in deep ocean water, *Science*, *147*, 1291, 1965.

Harris, J. E., Characterization of suspended matter in the Gulf of Mexico - I. Spatial distribution of suspended matter, *Deep Sea Res.*, *19*, 719-726, 1972.

Jacobs, M. B. and M. Ewing, Suspended particulate matter: Concentration in the major oceans, *Science*, *163*, 380-383, 1969.

Lisitzin, A. P., Distribution and composition of suspended material in the sea and ocean (in Russian), in *Proceedings of Conference on Recent Sedimentation in the sea and ocean*, Moscow, May 24-27, 1960, 175-232, 1961.

Lisitzin, A. P., Sedimentation in the world ocean, *Spec. Pub. 17*, 218 pp., SEPM, 1972.

Montgomery, R.B., Equatorial undercurrent observations in review, *J. Oceanog. Soc. of Japan, 20th Ann. vol.*, 487, 1962.

Ochakovsky, Y., On the dependence of the total attenuation coefficient upon suspensions in the sea, *U.S. Dept. Comm., Joint Publ. Ser. Rep. 36816, Mon. Cat. 13534*, 1966.

Pak, H., The Columbia River as a source of marine light scattering particles, Ph.D. Thesis, Oregon State University, Corvallis, 1970.

Pak, H., J. R. V. Zaneveld, and G. F. Beardsley, Jr., Mie scattering by suspended clay particles, *J. Geophys. Res.*, *76*, 5065, 1971.

Pak, H. and J. R. V. Zaneveld, The Cromwell Current on the east side of the Galapagos Islands, *J. Geophys. Res.*, *78*, 7845-7859, 1973.

Parsons, T. R., Suspended organic matter in sea water, in *Progress in Oceanography*, vol. 1, edited by M. Sears, pp. 205-232, Pergamon Press, London, 1963.

Plank, W. S., H. Pak, and J. R. V. Zaneveld, Light scattering and suspended matter in nepheloid layers, *J. Geophys. Res.*, *77*, 1689, 1972.

Plank, W.S., J.R.V. Zaneveld, and H. Pak, The distribution of suspended matter in the Panama Basin, *J. Geophys. Res.*, *78*, 7113-7121, 1973.

Sheldon, R. W. and T. R. Parsons, *A practical manual on the use of the Coulter Counter in marine research*, Coulter Electronic Co., Toronto, 1967.

Sheldon, R.W., A. Parkash, and W.H. Sutcliffe, Jr., The size distribution of particles in the ocean, *Limnol. Oceanog.*, *17*, 327, 1972.

Spilhaus, A. F., Jr., Observations of light scattering in sea water, ph.D. Thesis, Massachusetts Institute of Technology, Cambridge, Mass., 1965.

Van Andel, T. J., G. R. Heath, B. T. Malfait, D. F. Heinrichs, and J. I. Ewing, Tectonics of the Panama Basin, Eastern Equatorial Pacific, *Geol. Soc. Am. Bull.*, *82*, 1489, 1971.

Zaneveld, J. R. V., M. Andrade, and G. F. Beardsley, Jr., Measurements of optical properties at an oceanic front observed near the Galapagos Islands, *J. Geophys. Res.*, *74*, 5540, 1969.

Zaneveld, J.R.V., and H. Pak, Method for the determination of the index of refraction of particles suspended in the ocean, *J. Opt. Soc. Amer.*, *63*, 321, 1973.

Suspended Matter and the Stability of the Water Column: Central Caribbean Sea

N. JAY BASSIN

Texas A & M University

ABSTRACT

Total suspended matter, TSM, was measured in the water column at 4 stations in the central Caribbean Sea in relationship to the vertical profile of static stability. It is postulated that the several TSM maxima at depth are associated with zones of low turbulence, and the near-bottom nepheloid layer observed in many areas of the world ocean is a function of a near-bottom density increase coupled with a bottom shear capable of suspending fine sediment. The absence of one or both of these criteria explains the reported lack of a nepheloid layer in the Caribbean.

INTRODUCTION

This paper is the third in a series characterizing the total suspended matter, *TSM*, in the Caribbean Sea. *Bassin et al.* [1972] discussed the vertical distribution of *TSM* at several stations (including the 4 presently reported) by means of gravimetric analysis. They pointed out the lack of evidence in the Caribbean for a near-bottom nepheloid layer as reported in other ocean basins [*Ewing and Thorndike*, 1965; *Eittreim et al.*, 1969; *Hunkins et al.*, 1969; *Ewing and Connary*, 1970; *Plank et al.*, 1972]. Nepheloid layers are defined on the basis of optical, or light-scattering, data and the lack of a zone of increased light-scattering has been noted by other workers in the Caribbean [*Betzer and Pilson*, 1971; *Feely and Sullivan*, 1972]. Another widely occurring aspect of vertical *TSM* distribution is the existence of a mid-depth maximum, or succession of maxima, in the water column. *Jerlov* [1959], using optical data, related the mid-depth maxima to eddy diffusivity while *Lisitzin* [1972] reported good correlation between the depths of the maxima and nearby sills.

271

Ichiye et al. [1972] showed good agreement between location of mid-depth maxima, vertical eddy diffusivity computed from gravimetric data, horizontal advection of *TSM* from sills bordering the northern Caribbean, and eddy diffusivity derived from hydrographic data. They interpreted the maximum in the northern Caribbean as scour from the sill-depth of Mona Passage.

Bouma et al. [1969] point out that horizontal transport of suspended material occurs at density interfaces within the water column, yet particle maxima are not always associated with such pycnoclines. *Carder* [1970] concluded that the density (temperature and salinity) of the water mass does not correlate with particle size; and *Carder et al.* [1971] plot light-scattering data with the Brunt-Väisällä frequency to show a correlation at the pycnocline region. No relationship has yet been made between density stratification and gravimetric *TSM* concentration in the open ocean.

The purpose of this paper is to show that a qualitative relationship may be recognized between the vertical distribution of *TSM* and the distribution of stability in the water column.

METHODS

Total suspended matter was determined from 4 stations in the central Caribbean (Figure 1) at several depths by gravimetric analysis. Briefly described, sea-water samples were collected using 5- and 30-ℓ Niskin bottles, together with Nansen bottles which were utilized for hydrographic data. The samples were pressure-filtered through

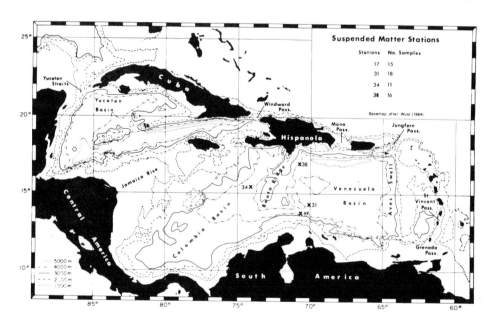

Fig. 1. Location map of suspended-matter stations collected in the Caribbean Sea during cruise 71-A-4 of the R/V *Alaminos*.

0.45 μm Millipore filters which had been prepared and tared by the control-filter technique described in *Bassin et al.* [1972] and *Harris* [1972]. The soluble organic wetting agent applied to the filter pads at the factory was removed before taring by two washings in large volumes of distilled water. Residual salts were rinsed from the wet filter pads by several washings of double-distilled deionized water (approximately 100 cc) until the rinse water showed no trace of Cl_2 when tested with 1% $AgNO_3$ solution. This rinsing must be done immediately after sample filtration; otherwise not all the salt may be removed. The rinsed pads were stored over NaOH pellets in a desiccator until they could be reweighed.

Analysis of variance showed that the weighing system used is accurate to ±27 μg at a 95% confidence level. Since the volumes filtered varied between 5 and 20ℓ, the residual error on the *TSM* concentration varied between ±5.5 and ±1.5 μg/ℓ.

The hydrographic data was obtained from Nansen bottles lowered on the same wire used for the Niskin bottles. Salinity was determined by standard titration techniques on-board ship and temperature was read off reversing thermometers.

STABILITY

Static stability, E', is a measure of the resistance of a column of water to overturning, and is expressed as a gradient of density. *Hesselberg and Sverdrup* [1915] defined it by

$$E' = 10^3 \ (d\sigma\theta/dz) \qquad (1)$$

in units of gm cm^{-3} m^{-1}, and this equation was used to calculate the stabilities plotted in Figure 2. It is recognized that the limited amount of data presented here is insufficient to quantitatively characterize these correlations, and this model may represent only broad-scale relationships between *TSM* and stability. A more rigorous model of the hypothesis must await a solution to the problem of hydrodynamic behavior of a two-phase system (water and particles) at a density interface, as well as much closer sampling intervals together with continuous STD measurements and nephelometry.

Qualitatively, the intriguing aspect of the four comparisons is that the various maxima of *TSM* are usually above the associated E'-maxima. A similar relationship may be noted in the plot of Brunt-Väisällä frequency with light scattering presented by *Carder et al.* [1971]. The explanation for this phenomenon lies in the fact that the stability maxima are regions of minimum advective vertical transport, and thus behave as traps for descending particles. Even when the stability maximal layer is a laminar layer, the regions above and below may be turbulent. This turbulence is required to maintain a particle-concentration gradient above each interface and — in the simplest case of a two-layer ocean with particle influx from above — would produce the stability/particle concentration

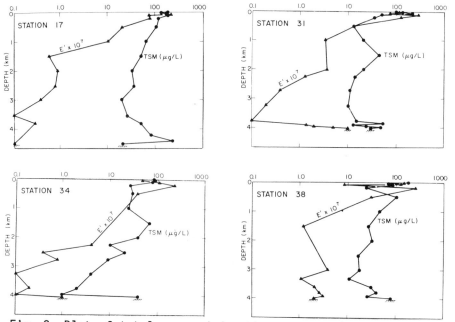

Fig. 2. Plot of total suspended matter, *TSM*, and static stability
E'.

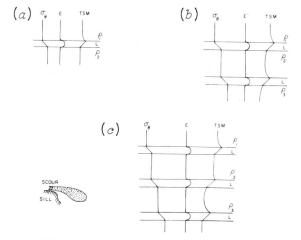

Fig. 3. Idealized layered ocean with laminar boundaries, *L*, and
TSM input from the surface. (*a*) Two-layered ocean; (*b*) Multilayered
ocean; (*c*) Multilayered ocean with submarine particulate source in
addition to surface influx.

relationship shown in Figure 3*a*. The maximum *TSM* may occur above
the "laminar layer" (high stability) where turbulence is minimal.
It follows, therefore, that the hypothetical relationships shown in
Figure 3 depend on a nearly laminar boundary layer. The criterion
for laminar motion is represented by the dimensionless Richardson

number, Ri, which is the ratio of the static stability (potential energy) to the velocity shear (du/dz; available kinetic energy). The intensity of vertical turbulence varies inversely with Ri.

$$Ri = g \ E' \ (du/dz)^{-2} \tag{2}$$

It is therefore possible to determine the layer of maximum stability by calculating the Richardson number. If a critical Ri of 0.52 [*Bowden*, 1962] is assumed, the minimum static stability becomes 0.52 x 10^{-7} for a maximum current shear of 30 cm $sec^{-1}/1000$ m in the 1-2 km range [*Ichiye and Sudo*, 1971]. This threshold is comparable to the minimum static stability values observed between 1 and 2 km depth (Figure 2). Since the intensity of vertical turbulence varies inversely with the stability for a constant current shear, the suspended matter will be more dispersed vertically both above and below the level of stability maxima than in that layer. It is postulated that the maximum concentration of *TSM* increases above a layer of E' maximum, which is almost laminar, since particles fall freely in a Stokesian manner through the layer until turbulence below retards them. Although turbulence advects both upwards and downwards, the "transit time" of a particle descending under gravity will be greater under turbulent conditions than under laminar conditions. Thus, the stability maximum remains freer of particles.

   Ichiye [1966a] reports that the increment of density increase due to *TSM* is on the order of 10^{-6} to 10^{-4} gm/cm^3. To see if the increased density balances the "buoyancy" of the turbulent zones, combine (1) and (2), and obtain:

$$Ri = 10^3 \ g \ \frac{d\sigma\tau}{dz} \ \frac{du}{dz}^{-2} \tag{3}$$

where $\sigma\tau$ represents the total density (water + *TSM*) in Sigma-t units (10^{-3} g/cm^3). The total density gradient may be represented by

$$\frac{d\sigma\tau}{dz} = \frac{d\sigma_\omega\Theta}{dz} + \frac{d\sigma_s}{dz} \tag{4}$$

where $\sigma_\omega s$ and σ_s represent the densities of water and *TSM*, respectively. Let the change in Richardson number, ΔRi, represent the difference between the observed value of seawater stability (where $\sigma = \sigma\tau$) and that value which represents the stability without the suspended matter ($\sigma = \sigma_\omega s$):

$$\Delta Ri = 10^3 \ g \left(\frac{d\sigma\tau}{dz} - \frac{d\sigma_s}{dz} \right) \left(\frac{du}{dz} \right)^{-2} \tag{5a}$$

or
$$\Delta Ri = 10^3 \ g \left(\frac{d\sigma_\omega s}{dz} \right) \left(\frac{du}{dz} \right)^{-2} \tag{5b}$$

   In the Caribbean between 1 and 2 km depth, the $\sigma_\omega\Theta$ changes from approximately 26 to 27 [*Ichiye and Sudo*, 1971], while the amount of particulate matter drops by about 20 µg/ℓ (= 20 x 10^{-9} g/cm^3). Inserting these values in (5b), ΔRi becomes about 20 x 10^{-8}, in cgs

units. Even though this is only a rough estimate, it is evident
that the amount of stability induced by *TSM* in the Caribbean mid-
depths is negligible, and is on the order of 10^{-2} to 10^{-1} of the
potential density at that depth. Particles, after being retarded
by the overlying turbulence layer, will pass through the more stable
interface and may set up an equilibrium situation in which the flux
of particles passing through the boundary balances the flux of sediment being introduced from above. The detritus passing through the
layer will then build up over the next-deeper stability maximum,
until an idealized multilayer situation is realized (Figure 3*b*).
Bottom and surface boundary effects are not taken into consideration here, although it is assumed that particles are removed from
the system by sedimentation and horizontal advective transport and
that particles are introduced from the surface. Mid-depth addition
of sediment (as from horizontal scour from a submarine ridge or
sill) would simply increase the total *TSM* concentration between
maxima, without affecting its settling behavior, as depicted schematically in Figure 3*c*.

NEAR-BOTTOM NEPHELOID LAYER

It seems apparent that the maintenance of a permanent nepheloid
zone near the bottom is due to the presence of a laminar layer
(where *Ri* is larger than critical) because of the increase of a *TSM*-
induced density gradient [*Ichiye*, 1966*b*]. This *TSM* density component would have to be over the threshold limit necessary to decrease
ΔRi by at least two orders of magnitude. In the Caribbean, the
critical *TSM* concentration in the near-bottom layer would have to
be above about 2 mg/ℓ, clearly far above any reported near-bottom
value.

In equation 3, the near-bottom velocity gradient $(du/dz)^2$ is
larger because $\mu = 0$ at $z = 0$ (Figure 4) and the value $d\sigma_s/dz$ in-

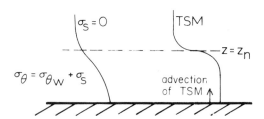

Fig. 4. Relationship of *TSM* to density at the bottom in a nepheloid zone. Due to the positive (downward) gradient of density, a
stable layer is formed, preventing sediment advecting upward from
penetrating the nonturbulent region ($z = z_n$).

crease upwards from the bottom due to vertical advection of suspended matter from the sediment [*Ichiye et al.*, 1972]. Thus,

$$Ri = 10^3 \, g \, \frac{\frac{d\sigma \Theta \omega}{dz} + \frac{d\sigma_s}{dz}}{(du/dz)^2} \qquad (6)$$

may exceed Ri_c (critical Richardson number) at some height from the bottom $z = z_n$, and create a laminar layer at that distance from the bottom. In this case, the *TSM* cannot move through the z_n layer and decreases sharply (Figure 4).

In the Caribbean Sea, a nepheloid layer does not exist as a permanent feature for the following reason: The vertical velocity gradient apparently drops off considerably near the bottom [*Ichiye and Sudo*, 1971], resulting in a higher Ri. This indicates that turbulence near the bottom is considerably reduced and thus the continuous suspension of particles from the bottom is prevented.

CONCLUSIONS

Vertical distribution of *TSM* in the ocean appears to be directly related to water mass through the stability function. Mid-depth *TSM* maxima occur in the Caribbean in regions overlying high stability layers and theoretical distributions conform to observed structure at 4 sampling stations. The absence of a permanent nepheloid layer in the Caribbean is explained by the high stability of the water column near the bottom, preventing turbulent diffusion upward from the sediment, as well as low, bottom-shear velocities. Considerably more work must be done on the stability problem, particularly in regions of known nepheloid layers. Future work should consider continuous vertical profiling by means of *in-situ* STD records and nephelometry in order to examine the small-scale effect of stability on suspended particulates as well as to rigorously quantify their relationship.

ACKNOWLEDGEMENTS

The writer is indebted to Professor T. Ichiye for his help and valuable suggestions on this paper. Portions of this work were done in partial fulfillment of graduate degree requirements. Financial assistance was made possible by Dr. A. H. Bouma under Office of Naval Research Contract N00014-68-A-0308-0002.

REFERENCES

Bassin, N. J., J. E. Harris, and A. H. Bouma, Suspended matter in the Caribbean Sea: a gravimetric analysis, *Mar. Geol.*, *12: 3*, M1-5, 1972.

Betzer, P. R., and M. E. Q. Pilson, Particulate iron and the nepheloid layer in the western North Atlantic, Caribbean and Gulf of Mexico, *Deep Sea Res.*, *18*, 753-761, 1971.

Bouma, A. H., R. Rezak, and F. B. Chmelik, Sediment transport along oceanic density interfaces, *Geol. Soc. Am.*, *Abs. Prog.*, *7*, 259-260, 1969.

Bowden, K. F., Turbulence, *The Sea*, vol. 1, edited by M. N. Hill, p. 802, Wiley Interscience, New York, 1962.

Carder, K. L., Particles in the Eastern Equatorial Pacific Ocean: their distribution and effect upon optical parameters, Unpublished Ph.D. thesis, Oregon State University, Corvallis, Oregon, 1970.

Carder, K. L., G. F. Beardsley, and H. Pak, Particle size distributions in the Eastern Equatorial Pacific, *J. Geophys. Res.*, *76*, 5070-5077, 1971.

Eittreim, S., M. Ewing, and E. M. Thorndike, Suspended matter along the continental margin of the North American Basin, *Deep Sea Res.*, *16*, 613-624, 1969.

Ewing, M., and S. D. Connary, Nepheloid layer in the North Pacific, *Geol. Soc. Amer. Mem.* *126*, 41-82, 1970.

Ewing, M., and E. Thorndike, Suspended matter in deep ocean water, *Science*, *147*, 1291-1294, 1965.

Feely, R. A., and L. Sullivan, Correlation between light scattering measurements and suspended material analysis in the characterization of the nepheloid layer, (abstract), *Am. Geophys. U.*, *53rd Annual Meeting*, *Trans. Am. Geophys. U.*, *53: 4*, 424, 1972.

Harris, J. E., Characterization of suspended matter in the Gulf of Mexico. I. Spatial distribution of suspended matter, *Deep Sea Res.*, *19*, 719-726, 1972.

Hesselberg, Th., and H. U. Sverdrup, Die stabilitatsverhaltnisse des seewassers bei vertikalen verschiebungen, *Bergens Museums Aarbok 1914-15*, *15*, (English transl., U.S. Navy Hydrographic Office, Trans. 148, 1-16, 1962), 1915.

Hunkins, K., E. M. Thorndike, and G. Mathieu, Nepheloid layers and bottom currents in the Arctic Ocean, *J. Geophys. Res.*, *74*, 6995-7008, 1969.

Ichiye, T., Some hydrodynamic problems for a nepheloid zone, *Pure App. Geophys.*, *63*, 179-195, 1966*a*.

Ichiye, T., Turbulent diffusion of suspended particles near the ocean bottom, *Deep Sea Res.*, *13*, 679-685, 1966*b*.

Ichiye, T., N. J. Bassin, and J. E. Harris, Diffusivity of suspended matter in the Caribbean Sea, *J. Geophys. Res.*, *77*, 6576-6588, 1972.

Ichiye, T., and H. Sudo, Saline deep water in the Caribbean Sea and in the Gulf of Mexico, *Tech. Rept. 71-16-T*, Department of Oceanography, Texas A&M University, College Station, Texas, July, 1971.

Jerlov, N. G., Maxima in the vertical distribution of particles in the sea, *Deep Sea Res.*, *5*, 173-184, 1959.

Lisitzin, A. P., *Sedimentation in the World Ocean, Spec. Pub. 17*, SEPM, Tulsa, Oklahoma, 218 pp., 1972.

Plank, W. S., H. Pak, and J. R. V. Zaneveld, Light scattering and suspended matter in nepheloid layers, *J. Geophys. Res., 77*, 1689-1694, 1972.

Light-Scattering Measurements and Chemical Analysis of Suspended Matter in the Near-Bottom Nepheloid Layer of the Gulf of Mexico

RICHARD A. FEELY, LAWRENCE SULLIVAN, AND
WILLIAM M. SACKETT

*Texas A&M University; Lamont-Doherty Geological
Observatory; Texas A&M University*

ABSTRACT

*The mass and chemical composition of suspended material in three
near-bottom profiles in the Gulf of Mexico and in one profile in the
Caribbean show a definite relationship to light scattering as measured by an Ewing-Thorndike nephelometer. For the Gulf of Mexico
profiles, an increase in light scattering from about 300-700 above
the bottom indicates a near-bottom nepheloid layer. In this layer,
at about 100-200 m above the bottom, the concentrations of particulate aluminum, silicon, iron and organic carbon generally increased
from 2 to 6 times over their respective concentrations at mid-depth.
Constant and low light-scattering values and suspended matter concentrations at one station indicate that no near-bottom nepheloid
layer is found in the Yucatan Basin of the Caribbean.*

INTRODUCTION

In 1953, *N. G. Jerlov* published a paper entitled "Particle Distribution in the Ocean" in which he reported maxima in light scattering in specific regions throughout the seawater column and, in
some cases, adjacent to the ocean bottom at stations in the Indian
Ocean, Red Sea and the eastern and western basins of the Atlantic
Ocean. Jerlov attributed the maxima in light scattering to increased concentrations of suspended particulate matter. *Ewing and Thorndike* [1965] measured light scattering along the continental slope
and rise of the northwestern Atlantic and demonstrated that the
water in the bottom few hundred m generally scattered light several
times more than water at intermediate depths. They referred to
this region as the nepheloid layer.

In the 20 years since the pioneering work of Jerlov, the existence of a near-bottom nepheloid layer has been reported in the northeastern Atlantic [*Jerlov*, 1953; *Jones et al.*, 1970], northwestern Atlantic [*Eittreim et al.*, 1969], Arctic Ocean [*Hunkins et al.*, 1969], Argentine Basin [*Ewing et al.*, 1967; *Ewing et al.*, 1971], North Pacific [*Ewing and Connary*, 1970], eastern equatorial Pacific [*Pak et al.*, 1970], Indian-Pacific Antarctic Sea [*Eittreim et al.*, 1972], and the Gulf of Mexico [*Ewing et al.*, in preparation, 1973].

The basic premise upon which all these investigations were made is that the increase in light scattering is due to increases in particulate matter concentrations. *Jacobs and Ewing* [1969] centrifuged sea water from the major oceans and found concentrations of particulate matter averaging 50 μg/ℓ in deep water. For the North Atlantic, an increase in particulate matter concentrations averaging 70 μg/ℓ was found in the nepheloid layer compared with 49 μg/ℓ for the clear water above it. *Bassin et al.* [1972] found significant increases in suspended matter concentrations at 200 m above the bottom in the northwestern Caribbean Sea but were uncertain whether this increase represented a nepheloid layer. Furthermore, the only information available regarding the composition of the suspended matter in the nepheloid layer comes from two sources. *Ewing and Thorndike* [1965] filtered a large quantity of water from the nepheloid layer and identified the suspended material as "lutite". *Betzer and Pilson* [1971] filtered sea water from the nepheloid layer in the Gulf of Mexico, the Caribbean and the Atlantic and found increases in particulate iron concentrations of more than four times the average of the concentrations of particulate iron in the clear water above the nepheloid layer.

A feasibility study was initiated on cruise 71-A-12 of the R/V *Alaminos* during October, 1971 (in cooperation with Maurice Ewing) to determine the relationship between particulate matter concentrations and light-scattering measurements in the near-bottom nepheloid layer of the Gulf of Mexico. This paper presents the results of that study and should be considered as a preliminary survey. Comprehensive surveys of the near-bottom nepheloid layer in the Gulf of Mexico have been made on subsequent cruises on the R/V *Alaminos* and the R/V *Trident* and will be reported on in later papers.

SAMPLING AND ANALYTICAL PROCEDURES

Light-scattering measurements were taken on every hydrocast, using an *in situ* photographic nephelometer described by *Thorndike and Ewing* [1967]. Figure 1 shows the sampling system which consisted of the nephelometer, an acoustic transmitter (pinger), a Nansen bottle, eight 30-ℓ Niskin bottles and another Nansen bottle. Sea water samples from the 30-ℓ Niskin bottles were filtered under a pressure of 40 psi through preweighed 0.40-μm Nuclepore filters at station 3 and through preweighed and pretreated 0.45-μm Millipore filters at all other stations. This system was used to determine the relative merits of the two types of filters. The filters were

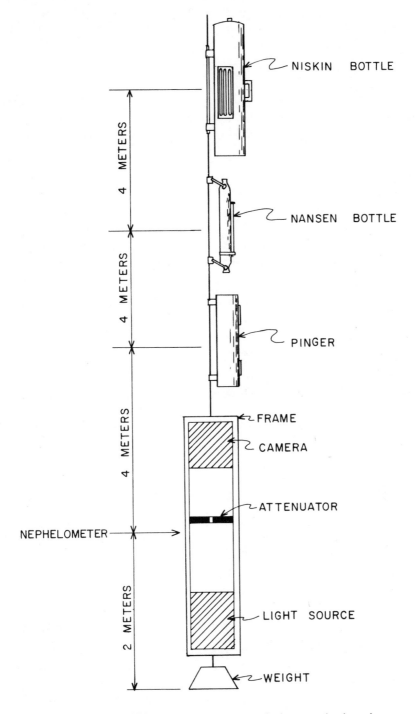

Fig. 1. The sampling system suspended on a hydrowire.

washed with 30 ml of triply distilled, filtered deionized water to
remove dissolved salts, dried in a desiccator, and stored in plastic
petri dishes for shipment to the laboratory.

In the laboratory, the filters were weighed to the nearest 0.01
mg on an Ainsworth Model 24N semi-micro balance according to the
method of *Banse et al.* [1963], folded, placed in a platinium cruci-
ble and ashed. The samples were then fused with 1.7 g reagent-grade
sodium carbonate, treated with 10 ml 6N hydrochloric acid and diluted to 25 ml with triply distilled, deionized water. Aliquots of
these samples were analyzed for particulate aluminum, silicon, and
iron. The aluminum content was analyzed by the fluorometric method
of *Sackett and Arrhenius* [1962] with modifications suggested by
Donaldson [1966]. Silicon was determined by the spectrophotometric
method of *Strickland and Parsons* [1968]. Iron was determined by a
method suggested by *Peter R. Betzer* [personal communication, 1972]
using a Perkin-Elmer HGA 2000 Heated Graphite Atomizer. One-ml
aliquots of the samples were introduced into a 100-ml Nalgene volumetric flask. One ml of concentrated hydrochloric acid was added
to the flask and the sample was diluted to 100 ml with triply distilled, deionized water. Twenty five aliquots of one-μl volume of
this sample were introduced into the graphite atomizer and subjected
to the following procedure which removed interferences by selective
volatilization:

(*a*)   drying temperature, 100°C; drying time, 10 sec
(*b*)   charring temperature, 1000°C; charring time, 20 sec
(*c*)   atomization temperature, 2500°C; atomization time, 10 sec.

Concentrations of particulate organic carbon were obtained separately. Two one-l aliquots from the 30-l Niskin bottle samplers were
filtered through precombusted 0.45-μm Gelman type A glass fiber filters. The filters were then analyzed for particulate organic carbon
by the wet-combustion infrared method of *Fredericks and Sackett*
[1970].

A statistical analysis of the errors involved with the determinations for aluminum, silicon and iron were evaluated by preparing
known amounts of W-1 standard rock samples from the U.S. Geological
Survey [*Fleischer*, 1969] in the same manner as the filtered samples.
The range of error was 1% to 10%, with an average error of 4% for
aluminum, 5% for silicon and 9% for iron.

RESULTS

Light-scattering measurements were combined with suspended material analyses for three stations in the Gulf of Mexico and one in
the Caribbean Sea (Figure 2 and Table 1). Light scattering was
measured on another station but time did not permit collection of
suspended matter.

Figure 3 shows a depth profile of station 3 located just off the
Texas-Louisiana Shelf. Total suspended matter (*TSM*), particulate or-
ganic carbon (*POC*), particulate aluminum (*PAl*), particulate silicon
(*PSi*), and particulate iron (*PFe*) were measured in eight near-bottom

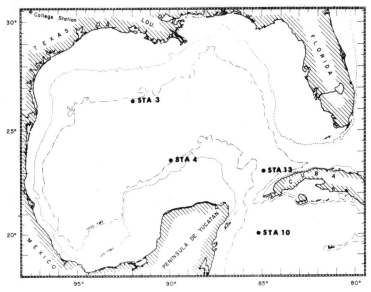

Fig. 2. The location of the stations on R/V *Alaminos* cruise
71-A-12.

samples along with light-scattering measurements. The light-
scattering measurements are reported in log scattering units (log S)
which represents the differences between the log of the exposure on
the film from the scattered light and the log of the exposure on the
film from the attenuated light. These differences compensate for
all variations in light source or film transport speed. Thus the
log S value is a relative value not dependent on the light source
or any other instrumental variable. The light-scattering data given
in Figure 3 indicates that the nepheloid layer begins about 500 m
above the bottom and increases to a maximum at about 200 m above the
bottom. *TSM* and *POC* concentrations begin to increase at about the
same depth and increase to a maximum at about 300 m above the bottom
for the *POC* and 100 m above the bottom for the *TSM*. At its maximum,
TSM is greater than 5 times the average of the values above the
nepheloid layer. The *PAl* and *PSi* concentrations begin to increase
about 200 m above the bottom and increase to a maximum at about 100
m above the bottom. *PAl*, at the maximum, is about 6 times the
average of the concentrations of *PAl* in clear water above the nepheloid layer; at the maximum, *PSi* is 7 times the average of the concentrations above the nepheloid layer. *PFe* concentrations (Table 1)
begin to increase about 150 m above the bottom and increase to a
maximum at 100 m above the bottom.
      Figure 4 shows the vertical profile of station 4 located in the
south central Gulf of Mexico. Light-scattering values show a gradual increase to the bottom starting at 700 m above the bottom. *POC*
concentrations show a maximum at 550 m above the bottom whereas *PAl*
and *PSi* show maxima at 65 and 15 m, respectively, above the bottom.
PFe concentrations show a maximum at about 100 m above the bottom.

Table 1. Suspended-Matter Analyses for Cruise 71-A-12 of the R/V *Alaminos*.

STATION	DEPTH (m)	POSITION	FILTER TYPE	TSM (µg/L)	POC (µg/L)	PAl (µg/L)	PSi (µg/L)	PFe (µg/L)
3	900	26°24.0N	Nuclepore	57	14.8	1.9	5.5	1.3
3	1400	91°56.0W	"	56	11.5	1.5	3.9	1.8
3	1600		"	66	15.4	1.0	4.2	0.5
3	1700		"	72	14.9	1.2	3.6	0.8
3	1750		"	226	13.8	5.8	18.8	1.8
3	1800		"	296	14.4	8.4	29.9	12.4
3	1850		"	68	15.2	1.3	4.8	0.2
3	1900		"	39	11.8	1.3	3.8	0.08
Bottom	1914							
4	2470	24°34.0N	Millipore	--	8.0	4.4	10.2	1.4
4	2970	89°48.0W	"	--	14.5	1.7	4.2	0.1
4	3170		"	--	9.6	1.5	3.7	0.1
4	3270		"	--	7.0	1.9	3.9	1.1
4	3320		"	--	8.9	1.1	3.8	1.1
4	3370		"	--	4.1	1.4	5.7	2.5
4	3420		"	--	5.7	2.6	5.8	2.0
4	3470		"	--	11.4	1.8	7.9	1.5
Bottom	3484							
10	2400	20°02.2N	"	--	5.2	1.1	2.8	0.5
10	3400	85°08.6W	"	--	7.6	0.8	3.2	0.2
10	3900		"	--	13.9	1.3	3.3	0.6
10	4200		"	--	3.6	0.5	3.2	0.6
10	4250		"	--	4.7	0.8	2.2	0.1
10	4300		"	--	5.0	0.8	1.9	0.08
10	4350		"	--	3.8	1.3	3.3	0.6
10	4400		"	--	4.0	0.6	3.1	0.09
Bottom	4414							
13	1020	23°00.9N	"	--	--	0.9	2.9	1.2
13	1520	84°55.1W	"	--	5.1	0.9	2.3	0.7
13	1720		"	--	--	0.7	2.1	0.6
13	1820		"	--	9.4	0.6	1.8	0.06
13	1870		"	--	10.5	0.9	2.2	0.2
13	1920		"	--	7.8	0.6	1.7	0.1
13	1970		"	--	--	4.0	9.0	1.2
13	2020		"	--	4.3	0.6	1.2	0.2
Bottom	2031							

TSM = total suspended matter; given for station 3 only, owing to erroneous *TSM* values obtained using Millipore filters due to incomplete removal of dissolved salts.
POC = particulate organic carbon
PAl = particulate aluminum
PSi = particulate silicon
PFe = particulate iron

Just as the light-scattering values show a gradual increase to the bottom so do the concentrations of *PSi* show a gradual increase to the bottom. *PAl* concentrations show no apparent trends with depth.
    Figure 5 shows the profile of station 10, located in the Yucatan Basin of the Caribbean Sea. The extremely low values of light scattering correspond with the low concentrations of *PAl*, *PSi* and *POC* relative to those of station 4. Average concentrations of *PAl*, *PSi* and *POC* are three times higher for *PAl*, two times higher for *PSi* and two times higher for *POC* in the deep water of the Gulf of Mexico (station 4) relative to the deep water of the Yucatan Basin. Except for a slight increase at about 100 m above the bottom, the concentrations of *PFe* remain low and constant with depth. There does not appear to be any evidence of a nepheloid layer at station 10 in the Yucatan Basin.

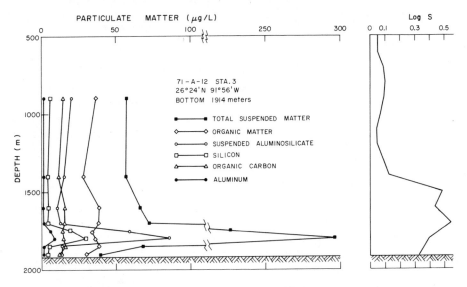

Fig. 3. Suspended-matter analyses and light-scattering measurements for station 3, cruise 71-A-12, R/V *Alaminos*.

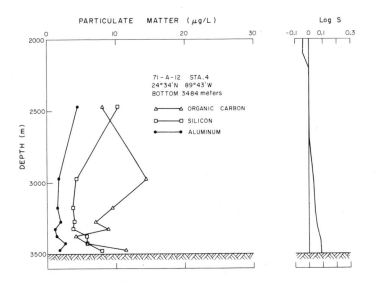

Fig. 4. Suspended-matter analyses and light-scattering measurements for station 4, cruise 71-A-12, R/V *Alaminos*.

Figure 6 shows the profile of station 13, located near the Florida Straits. Both light-scattering values and concentrations of *PAℓ*, *PSi* and *POC* are low and constant except for a region starting about 200 m above the bottom. Along with an increase in light scattering in this region, there is also a twofold increase in *POC*, a fourfold increase in *PAℓ* and a fourfold increase in *PSi*. *PFe* concentrations show an increase at about 60 m above the bottom.

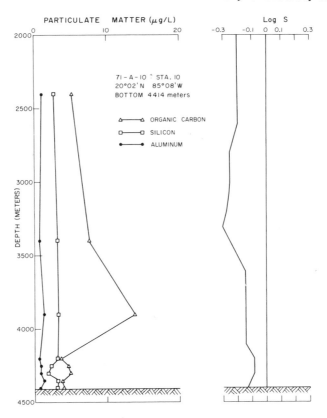

Fig. 5. Suspended-matter analyses and light-scattering measurements for station 10, cruise 71-A-12, R/V *Alaminos*.

DISCUSSION

Since most of the aluminum in the solid phase of pelagic sediments is bound in the crystalline lattice of aluminosilicate minerals, *Arrhenius* [1963] suggests that a rough estimate of the suspended aluminosilicate mineral concentration in sea water can be obtained from the particulate aliminum concentrations if the percent aluminum in the aluminosilicate fraction of the sediments is known. *Grim and Johns* [1954] reported the <1-μm fraction in two samples of sediments from the open Gulf of Mexico contains 10.06 and 10.13% aluminum by weight. If it is assumed that the <1-μm fraction of the Gulf of Mexico sediments reported by *Grim and Johns* [1954] represents aluminosilicate mineral grains, then a rough estimate of the suspended aluminosilicate concentrations can be obtained by multiplying the particulate aluminum concentration by a factor of 10.

Using this approach, a knowledge of the percentage of particulate organic carbon in particulate organic matter can be used to estimate particulate organic matter concentrations from measured particulate organic carbon concentrations in seawater. *Strickland*

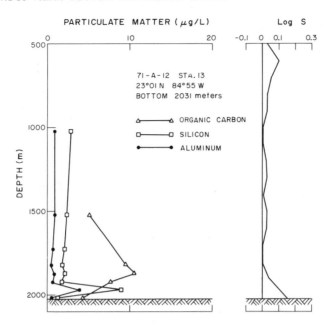

Fig. 6. Suspended-matter analyses and light-scattering measurements for station 13, cruise 71-A-12, R/V *Alaminos*.

[1965] reported the carbon content of marine organisms (the progenitor of particulate organic matter) is in the range of 45 to 55% by weight. *Gordon* [1970] uses the factor 1.8 for estimating particulate organic matter concentrations from particulate organic carbon concentrations obtained from samples taken in the North Atlantic. In a review of the literature on particulate organic matter in seawater, *Riley* [1970] suggests that a factor of 2.0 should be used.

In Figure 3 in the present study, the factor 2.0 is used to obtain the calculated concentrations of particulate organic matter (*POM*). Also in Figure 3, the factor 10 is used to calculate the concentrations of suspended aluminosilicates (*SAS*) which are presented for each of the samples in the profile. By subtracting the sum of the calculated amounts of *POM* and *SAS* from the *TSM* weights, the results can be used to obtain a value for the "aluminum-free inorganic suspended matter" [*Feely, Sackett and Harris*, 1971], which is due primarily to quartz, amorphous silica and the siliceous and calcareous skeletons of planktonic organisms. Although the siliceous and calcareous skeletons are biogenic in origin, they are inorganic from the point of view of a chemist. Thus, chemical methods can be used to quantitatively characterize suspended matter into its various organic and inorganic fractions. Figure 7 shows the separation of suspended matter into three main groups: suspended aluminosilicates (*SAS*), particulate organic matter (*POM*) and aluminum-free inorganic suspended matter (*AFISM*). Data for the surface and midwater samples from the Gulf of Mexico are from *Feely, Sackett and Harris* [1971]. At 10 m the suspended material is 4%

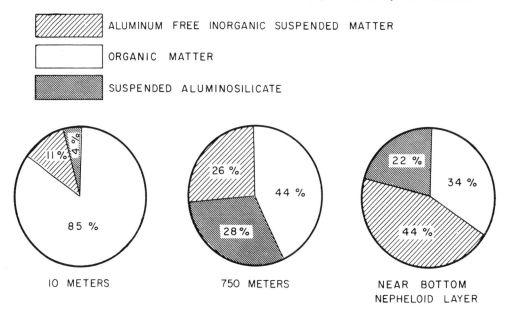

Fig. 7. Percentage of total suspended matter attributable to particulate organic matter (POM), suspended aluminosilicates (SAS), and aluminum-free inorganic suspended matter (AFISM) at various regions in the water column.

SAS, 85% POM and 26% AFISM. At mid-depth (the average of two samples, one at 500 m and one at 750 m) the suspended material is 30% SAS, 44% POM and 26% AFISM. At the near-bottom nepheloid region (an average of the last five samples of station 3), the suspended material is 22% SAS, 34% POM and 44% AFISM. The decrease in the percentage of particulate organic matter from surface to mid-depth is due to oxidation of organic matter. Considering the large errors involved in making a calculation such as this, there does not appear to be any significant variation between the mid-water and near-bottom nepheloid layer in so far as the composition of suspended matter is concerned. More data taken with better precision and more data taken at the same location are needed to delineate this problem properly.

When the data from station 10 in the Yucatan Basin are compared with the PFe concentrations obtained by Betzer and Pilson [1971] in the same region in April, 1969, during a cruise on the R/V *Trident*, an interesting situation occurs. Betzer and Pilson's *PFe* concentrations for station 13 (located approximately 90 miles ENE of station 10) show a sharp increase near the bottom, indicating the possible evidence of a nepheloid layer. But the suspended matter concentrations and light-scattering measurements taken 2 1/2 years later at station 10 and reported in the present paper show no evidence of a nepheloid layer. If the two sets of data are real, then either the bottom current velocities within the Yucatan Basin decrease rapidly from east to west (assuming nepheloid layers are produced by bottom currents) or the nepheloid layer in this region is episodic rather than static. In either case, it becomes necessary, when studying the nepheloid layer, to consider possible time-related variations.

ACKNOWLEDGEMENTS

The authors would like to express their appreciation to the following persons, without whose help this study would not have been possible. Dr. Maurice Ewing provided the nephelometer used for the study. Dr. Peter R. Betzer provided his laboratory facilities for particulate iron determinations. Mr. Alan D. Fredericks made the *POC* determinations. Financial support for the project was provided by the Office of Naval Research Contracts N00014-68-A-0308-0002 and N00014-67-0108-0004 and National Science Foundation Grant GA-27281. This paper is Lamont-Doherty Geological Observatory contribution 2047.

REFERENCES

Arrhenius, G., Pelagic sediments, *The Sea*, vol. 3, edited by M. N. Hill, pp. 55-718, Wiley Interscience, New York, 1963.
Banse, K. L., C. P. Falls, and L. A. Hobson, A gravimetric method for determining suspended matter in sea water using Millipore filters, *Deep Sea Res.*, *10*, 639-642, 1963.
Bassin, N. J., J. E. Harris and A. H. Bouma, Suspended matter in the Caribbean Sea: A gravimetric analysis, *Mar. Geol.*, *12*, 171-175, 1972.
Betzer, P. R. and M. E. Q. Pilson, Particulate iron and the nepheloid layer in the western North Atlantic, Caribbean, and Gulf of Mexico, *Deep Sea Res.*, *18*, 753-761, 1971.
Donaldson, D. E., Fluorometric analyses of the aluminum ion in natural waters, *Prof. Pap. 550D*, 258-261, U.S. Geol. Surv., Washington, D.C., 1966.
Eittreim, S., M. Ewing and E. M. Thorndike, Suspended matter along the continental margin of the North American Basin, *Deep Sea Res.*, *16*, 613-624, 1969.
Eittreim, S., A. L. Gordon, M. Ewing, E. M. Thorndike and P. Bruchhausen, The nepheloid layer and observed bottom currents in the Indian-Pacific Antarctic Sea, in *Studies in Physical Oceanography*, pp. 19-36, edited by A. L. Gordon, Gordon and Breach, London, 1972.

Ewing, M. and E. M. Thorndike, Suspended matter in deep ocean water, *Science*, *147*, 1291-1294, 1965.

Ewing, M., J. Ewing and X. LePichon, Sediment transport and distribution in the Argentine Basin, 4, Nepheloid layer and bottom water circulation (abstract), *Trans. Am. Geophys. Un.*, *48*, 141, 1967.

Ewing, M. and S. Connary, Nepheloid layer in the North Pacific, in *Geological Investigation of the North Pacific*, edited by J. Hays, pp. 41-82, *Geol. Soc. Amer. Mem. 126*, 1970.

Ewing, M., S. L. Eittreim, J. Ewing and X. Le Pichon, Sediment transport and distribution in the Argentine Basin, 3, Nepheloid layer and processes of sedimentation, in *Physics and Chemistry of the Earth*, *8*, pp. 49-77, Pergamon Press, New York, 1971.

Ewing, M., E. M. Thorndike and L. Sullivan, Nepheloid layer in the Gulf of Mexico, in preparation, 1973.

Feely, R. A., W. M. Sackett and J. E. Harris, Distribution of particulate aluminum in the Gulf of Mexico, *J. Geophys. Res.*, *76*, 5893-5902, 1971.

Fleischer, M., U.S. Geological Survey Standards: I. Additional data on rocks G-1 and W-1, *Geochim. et Cosmochim. Acta*, *33*, 65-69, 1969.

Fredericks, A. D., and W. M. Sackett, Organic carbon in the Gulf of Mexico, *J. Geophys. Res.*, *75*, 2199-2206, 1970.

Gordon, D. C., Some studies on the distribution and composition of particulate organic carbon in the North Atlantic Ocean, *Deep Sea Res.*, *17*, 233-244, 1970.

Grim, R. E., and W. D. Johns, Clay mineral investigation of sediments in the northern Gulf of Mexico, in *Clays and Clay Minerals, Proceedings of the Second National Conference*, 81-102, 1954.

Hunkins, K. E., E. M. Thorndike and G. Mathieu, Nepheloid layer and bottom currents in the Arctic Ocean, *J. Geophys. Res.*, *74*, 6995-7008, 1969.

Jacobs, M. B., and M. Ewing, Suspended particulate matter: Concentration in the major oceans, *Science*, *163*, 380-383, 1969.

Jerlov, N. G., Particle distribution in the ocean, in *Rep. Swedish Deep Sea Exped.*, *3*, 73-97, 1953.

Jones, E. J. W., M. Ewing, J. I. Ewing, and S. L. Eittreim, Influences of Norwegian Sea overflow water on the sedimentation in the northern Atlantic and Labrador Sea, *J. Geophys. Res.*, *75*, 1655-1680, 1970.

Pak, H., G. F. Beardsley, Jr. and W. Plank, Near-bottom nepheloid layer in the eastern Equatorial Pacific Ocean, *The Ocean World Symposium, Abstracts of Contributed Papers*, Tokyo, Japan, Sept. 13-25, 1970, p. 74.

Riley, G. A., Particulate organic matter in sea water, in *Adv. in Mar. Biol.*, *8*, pp. 1-118, Academic Press, New York, 1970.

Sackett, W. M. and G. Arrhenius, Distribution of aluminum species in the hydrosphere, 1. Aluminum in the oceans, *Geochim. et Cosmochim. Acta*, *26*, 955-968, 1962.

Strickland, J. D. H., Production of organic matter in the primary stages of the marine food chain, edited by J. P. Riley and G. Skirrow, in *Chemical Oceanography*, vol. 1, Chap. 12, Academic Press, London, 1965.

Strickland, J. D. H. and T. R. Parsons, *A Practical Handbook of Sea-
    water Analysis, Fisheries Res. Board of Canada, Bull. 167*, edited
    by J. D. Stevenson, 1968.
Thorndike, E. M. and M. Ewing, Photographic nephelometers for the
    deep sea, in *Deep Sea Photography*, pp. 113-116, John Hopkins Uni-
    versity Press, Baltimore, 1967.

Light Scattering and Suspended Particulate Matter on a Transect of the Atlantic Ocean at 11° N

PETER R. BETZER, KENDALL L. CARDER, AND
DONALD W. EGGIMANN

University of South Florida

ABSTRACT

A combined investigation of the optical, physical and chemical properties was carried out on the suspended particulate matter on a transect of the North Atlantic Ocean at about 11°N. A nepheloid layer approximately 100 m thick was found at near-bottom in the western basin of the North Atlantic; no evidence of a near-bottom nepheloid layer was found in the eastern basin. This interbasin difference is evident in measurements of suspended particulate matter and absolute light scattering β (45) and is thought to be due to differences in bottom water movement between the two basins. Within the western basin, near-bottom maxima in the above parameters coincides with the maximum flux of Antarctic Bottom Water. Increases in suspended particulate matter and absolute light scattering occur between 3000 and 4000 m at 2 stations near South America. Potential temperature/salinity characteristics of this water are consistent with its being a southerly extension of the Western Boundary Undercurrent. At 2 stations over the Mid-Atlantic Ridge, large (30-40%) increases in the mass of suspended particulate matter occur 30 m above the bottom. Particulate organic carbon and particulate carbonate determinations indicate these increases are due to greater amounts of refractory oxides and/or hydroxides in the suspended matter and may represent an injection or diffusion of materials from the ridge to deep ocean waters. Samples taken in Subtropic Underwater, oxygen minimum water and Antarctic Intermediate Water show relatively low amounts of suspended particulate matter and absolute light scattering, indicating that the nepheloid character reported for these water masses in the northwestern Caribbean and Yucatan Channel must be acquired in transit through the Carribbean Sea. Near-bottom nepheloid layers were found at four stations

*along a 240 km stretch of the African continental rise. The source
of much of the particulate matter between 1000 m and the bottom in
this part of the eastern basin is thought to be terrigenous material
delivered by a northwest-moving bottom current from the continental
shelves of Liberia and Sierra Leone.*

*Chemical analyses of samples from the mixed layer at eleven sta-
tions along the transect indicate atmospheric dust is the most likely
source for much of the near-surface suspended matter.*

INTRODUCTION

The general existence of nepheloid layers in deep, intermediate
and shallow waters of the world oceans is well established [*Hunkins
et al.*, 1969; *Eittreim*, 1970; *Ewing and Connary*, 1970; *Betzer and
Pilson*, 1971; *Eittreim et al.*, 1972]; but the sources, chemical and
mineralogic composition, and fluxes of suspended sediment in these
layers have yet to be determined. Our purpose here is to describe
the optical, physical, and chemical characteristics of the nepheloid layers found on a transect of the North Atlantic Ocean at 11°N.
In addition, the relationship of these layers to bottom water movement is considered and the sources of the suspended matter and how
the layers might be identified through chemical and mineralogic
analyses of the suspended matter is discussed.

MATERIALS AND METHODS

Water samples were collected in 5- and 30-ℓ Niskin bottles at
24 stations between Barbados, West Indies and Dakar, Senegal from
the R/V *Trident* (cruise 111) (Figure 1). Normally eight 30-ℓ and
fourteen 5-ℓ samples were collected at both shallow (surface-to-
1000 m) and deep (1000 m-to-bottom) stations. On the deep hydro-
casts, a pinger was used to locate the near-bottom samples which
were taken 30, 100 and 200 m from the bottom. The order of the
30-ℓ bottles on each hydrocast was assigned using a random number
table. In all, 380 samples were examined for light scattering, 214
samples were collected to study suspended particulate matter by
Coulter counter and for gravimetric procedures, and 210 were collected for use in determination of particulate organic carbon. Analyses for particulate carbonates were carried out on the 214 samples used in gravimetric determinations of suspended particulate
matter. Samples were drawn from all of the Niskin bottles for determination of salinity using an induction salinometer (Beckman
Model RS-7B). Water temperatures were collected using reversing
thermometers attached to the 5-ℓ Niskin bottles.

Continuous temperature/depth profiles were obtained at all the
shallow hydrographic stations (#1, 2, 4, 6, 8, 11, 12, 13, 15, 17,
18) using expendable bathythermographs (Sippican Model T4). Studies of suspended particles were carried out using a Brice-Phoenix
light-scattering photometer (Model 2000) described by *Spilhaus*

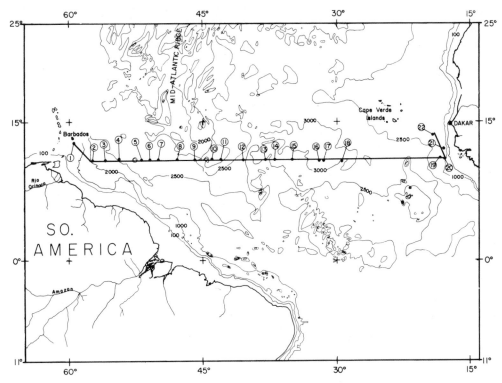

Fig. 1. Cruise track and station locations for the R/V *Trident* cruise 111 (depth contours are in fathoms).

[1965], *Pak* [1970])and a Coulter counter Model B (described by *Sheldon and Parsons* [1967]) equipped with a 100-μm diameter aperture. Based upon error estimates of instrument calibration and measurement repeatability by *Beardsley* [1966] and *Pak* [1970], respectively, *Carder* [1970] estimated the errors in absolute Brice-Phoenix measurements for $\beta(45)$, $\beta(90)$, and $\beta(135)$ to be less than 6% for each. *Carder* [1970] also estimated the error involved in measuring mean particle diameters (spherical equivalent) of marine particles to be less than 6%, including calibration and measurement repeatability errors using a Model A Coulter counter.

Processing of water samples, begun as soon as they were aboard ship, was completed within 3 hours of collection. After a portion of each sample had been removed from the 30-ℓ Niskin bottles for light scattering and Coulter counter study, the remainder of the water was gravity-filtered through 47-mm diameter Nuclepore 0.4μ pore-size filters and glass fiber filters (Type A). Filtration was accomplished in a closed system: the water was passed through a silicone rubber tube to an all-plastic, in-line filter head (Millipore Filter Corp.). Normally, about 18 ℓ of each sample was passed through a Nuclepore membrane and the remaining 8 ℓ through a glass-fiber filter. All filters were handled with teflon tweezers (separate pairs were used to handle each type of filter). After

filtration, all filters were placed in plastic Falcon tubes until
processing could be carried out in a shore-based "clean" laboratory.
The Nuclepore filters were rinsed twice in a clean bench (Baker
Model 3220) with approximately 4 mℓ doubly deionized water for each
rinse. Tests of the rinsing procedure show that, after the filters
were rinsed twice with deionized water, a small amount of residual
sea salt is retained by the particles and/or Nuclepore filter pad.
The 10 μg of salt retained introduces an average error of about
0.6 μg/ℓ into the data on suspended particulate matter presented in
this paper. After the rinse water had dripped through the filters,
each was placed in an acid-rinsed plastic tube, transferred to its
own polycarbonate bottle containing silica gel, dried for two days
and reweighed on a five-place Mettler balance (Model H20). Reweighing was accomplished in an air-conditioned room with a filtered
air supply, in order to reduce possibility of contamination. The
balance contained an *alpha* source to reduce static electric effects on
filter weights [*Manheim et al.*, 1970]. The weight of suspended particles on each Nuclepore filter was calculated by subtracting the
tare weight (each had been weighed before the cruise on the same
balance) from its desiccated weight and dividing by the volume of
water which had been filtered. While the precision of the weighing
procedure, as determined by replicate sample weighing, is equivalent
to ±0.5 μg/ℓ, a more realistic assessment of the uncertainties —
small variations in retention of sea salt, small variations in the
effects of desiccation under silica gel on weights of filter pads,
and small variations in the effects of sea water on the weights of
Nuclepore filters — associated with the data presented in Figures
2 and 6 is ±1.0 μg/ℓ.

Determinations of particulate organic carbon (*POC*) were made
using an infra-red analyzer (Beckman Model 215A) according to the
technique of *Menzel and Vaccaro* [1964]. The filters were oxidized
in a Coleman Carbon-Hydrogen Analyzer (Model 33) at 690°C to prevent
release of CO_2 from aragonite or calcite particles [*Fournier*, 1968].
No CO_2 was detected when samples of either aragonite or calcite were
placed in the carbon-hydrogen analyzer and run through the *POC* procedure at 690°C. The recovery of 200-μg portions of acetanilide run
through the *POC* procedure was 99 ± 0.9 percent for 7 samples. The
major uncertainty associated with the determination of *POC* content
of suspended materials was, however, the uncertainty of the carbon
content of the 47-mm glass-fiber filter pads. For an average volume
of 8 ℓ of filtered water, a more accurate estimate of the precision
of the *POC* procedure — taking into account the uncertainty of the
filter's carbon content — was ±1 μg/ℓ.

Determinations of particulate carbonate (*PC*) were made using a
modification of the acetic acid dissolution technique of *Chester and
Hughes* [1967]. In the present study, 4 mℓ of 25% (v/v) acetic acid
were added to each Nuclepore membrane (supported in a plastic funnel
having a teflon stopcock at its base) and left for 2 hours. Following the acid treatment each filter pad was rinsed twice with 4 mℓ
doubly deionized water for each rinse. The carbonate digestion procedure was checked using filtered samples of oceanic particulate

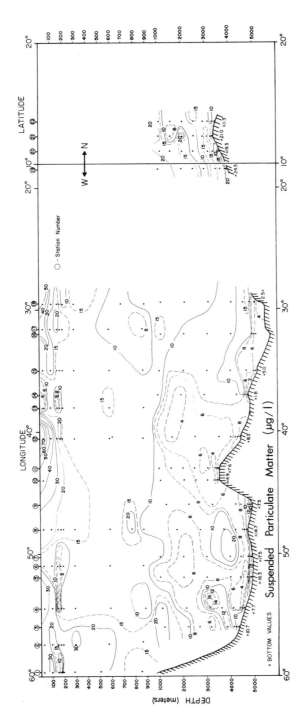

Fig. 2. Contours showing depth distribution of suspended particulate matter (μg/ℓ) for a transect of the North Atlantic Ocean at 11°N.

matter from the Gulf of Mexico. The Nuclepore membranes on which the
samples were collected were examined by microscope before and after
treatment with acetic acid. The numerous coccoliths present in these
samples were not detectable after exposure to acetic acid for 2 hours.
    The solutions were analyzed for calcium and magnesium concentra-
tions by atomic absorption spectrophotometry using a Perkin-Elmer
Model 403 equipped with a laminar flow burner and an air-acetylene
flame. Samples were interspersed with calcium and magnesium standards and run in quadruplicate. In order to minimize viscosity
effects on sample aspiration rates, the calcium and magnesium standards were prepared using the same acetic acid concentration that was
used on the samples. Blank corrections to the calcium and magnesium
determinations were made by leaching a Nuclepore filter with 25%
(v/v) acetic acid for 2 hours and then analyzing the solution for
its calcium and magnesium concentrations. The blank corrections
were extremely small, averaging 0.6 µg and 0.05 µg for calcium and
magnesium, respectively. The low calcium and magnesium levels in
the acetic acid resulted in average signal-noise ratios of 12:1 and
9:1 for calcium and magnesium, respectively.
    Calculations of PC concentrations for each sample were made by
multiplying the concentration of the particulate calcium by 2.5.
This procedure assumes that the particulate carbonate is a pure
$CaCO_3$; the high Ca/Mg ratios (10:1) from the samples studied indicate that this assumption is not unreasonable.

RESULTS AND DISCUSSION

    The distributions of suspended particulate matter and light
scattering $\beta(45)$ with depth for the transect of the North Atlantic
Ocean studied are presented in Figures 2 and 3, respectively. A
near-bottom nepheloid layer about 100 m thick is present in the
western basin (stations 3, 4, 5, 7, 8 and 9). In the eastern basin
(stations 12, 14, 15 and 18), suspended particulate matter and
light scattering decrease rather than increasing near the bottom.
Figure 4 summarizes the suspended matter and light-scattering data
for the two basins at 30, 100 and 200 m above the bottom. At 30 m
above the bottom, an average suspended load of 12.4 µg/ℓ was found
in the western basin whereas the average was 4.4 µg/ℓ in the eastern
basin. The difference between these two means is significant at the
99% confidence level. A comparison of samples taken 100 m above the
bottom also shows significant differences between the two basins.
The average suspended particle load in the western basin was
9.9 µg/ℓ and 5.5 µg/ℓ in the eastern basin. The difference between
these means is significant at the 95% confidence level. At 200 m
above the bottom, the average suspended load in the western basin is
6.0 µg/ℓ and 7.9 µg/ℓ in the eastern basin, a difference not considered significant at the 90% confidence level. The same trend is
also seen, however, in an interbasin comparison of light scattering
at 200 m above the bottom: $\beta(45)$ averaging 2.99 in the western
basin and 3.94 in the eastern basin. The relative increase in $\beta(45)$

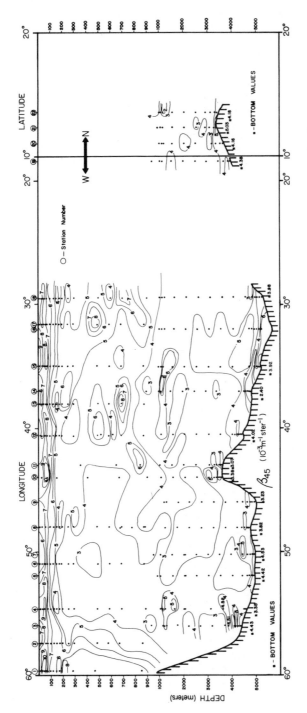

Fig. 3. Contours showing depth distribution of light scattering β(45) for a transect of the North Atlantic Ocean at 11°N.

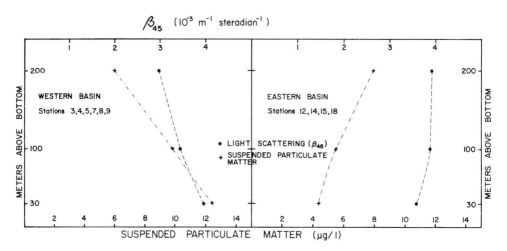

Fig. 4. Comparison of suspended particulate matter (μg/ℓ) and light scattering β(45) at 30, 100 and 200 m above the bottom in the western and eastern basins of the North Atlantic Ocean at 11°N.

compared to suspended particulate matter in the eastern basin will be commented upon later.

The development of a near-bottom nepheloid layer in the western basin of the North Atlantic Ocean can probably be related to the significant northwestward flux of Antarctic Bottom Water through this area. The Antarctic Bottom Water does not have access to the eastern basin (except in an insignificant sense through the Romanche Trench) since it is blocked by the Walvis and Mid-Atlantic Ridges [*Worthington and Metcalf*, 1961]. The low flux precludes significant interaction with bottom sediment of the eastern basin. The eastern basin is filled with North Atlantic Deep Water [*Wright*, 1970] and its deep water is not as active as the Antarctic Bottom Water is in the western basin [*Metcalf*, 1969].

Within the western basin significant trends in suspended particulate matter are also evident. At stations 3 and 4 on the western side of the basin average suspended loads of 9.2 and 7.2 μg/ℓ were found at 30 and 100 m, respectively, above the sediment/water interface. Stations 5, 7, 8 and 9 on the eastern side of the basin have corresponding averages of 14.0 and 11.2 μg/ℓ at these respective depths. *Wright* [1970] calculated the northward flux of Antarctic Bottom Water between 16°S and 16°N as a function of longitude in the western Atlantic Ocean. His calculations show that the maximum flux of Antarctic Bottom Water shifts from the South American (or western) side toward the Mid-Atlantic Ridge (or eastern) side of the western basin after this water mass crosses the equator. The fluxes calculated by Wright for Antarctic Bottom Water at 8°N are presented in Figure 5 (longitudes have been shifted slightly to match the actual

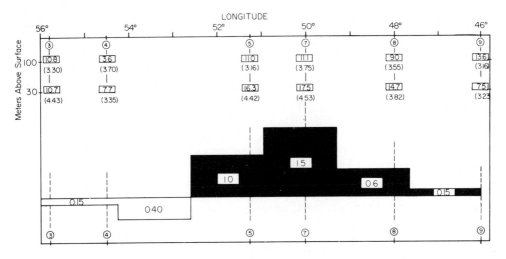

Fig. 5. Volume transport of Antarctic Bottom Water, suspended particulate matter (μg/ℓ) and light scattering β(45) as a function of
longitude in the western basin of the North Atlantic Ocean. Black
= northward transport; white = southward transport.

boundaries of the western basin at the latitude surveyed in the present study) along with data for the near-bottom suspended particulate
matter and light scattering for stations taken in the western basin.
On the western side of the basin only two stations were occupied and
both have the same flux of Antarctic Bottom Water, according to
Wright's calculations. On the eastern side of the basin, however,
samples were obtained from four different flux regions. From these
it can be seen that the maximum in suspended matter and light scattering corresponds with the maximum flux of Antarctic Bottom Water.
In fact, the amounts of suspended particulate matter and the magnitude of β(45) at 30 m above the bottom at stations 5, 7, 8 and 9
are distributed just as Wright's estimates of the bottom-water flux
indicated. Such a relationship implies the amount of suspended
particulate matter in the near-bottom nepheloid layer is a function
of the flux of bottom water over bottom sediments. *Wright* [1970]
is careful to point out that his calculations do not take into account any boundary currents which one would predict theoretically
to be operative near the South American side of the western basin
north of the equator. The particulate values at stations 3 and 4
may be related to this North Atlantic source instead. It is also
possible, however, that inputs from turbidity currents or the Amazon
cone are complicating the situation in the western portion of the
western basin.
    At stations 5, 7, 8, and 9, 23 samples of suspended particulate
matter collected between 780 and 4775 m gave an average of 8.1 μg/ℓ
compared to an average of 12.6 μg/ℓ for 8 samples collected within

110 m of the bottom at the same locations. The suspended matter in
this near-bottom nepheloid layer in the western basin has a markedly
different composition from that in overlying waters. *POC* contributes less to the materials in this layer than to the samples from
overlying waters: 22 samples collected between 780 and 4775 m gave
an average of 70% *POC*; the 8 samples collected 30 and 100 m above
the bottom gave an average of 43% *POC*. The contribution of carbonates to suspended matter increased slightly in the nepheloid layer:
24 samples collected between 780 and 4775 m gave an average of 3.3%
carbonate; the 8 near-bottom samples gave an average of 4.9% carbonate. Refractory materials such as clay minerals and quartz gave
an average of 27% of the suspended particulate matter in deep waters
above the nepheloid layer compared to 52% within it. These chemical
data show, then, that the 55% increase in suspended particulate matter in the 110 m of the bottom of the western basin is almost wholly
due to increases in refractory materials.

Ewing et al. [1967] reported finding a 1000 m-thick nepheloid
layer in the Argentine Basin in the western basin of the South Atlantic. Results for the nepheloid layer at 11°N show that, by the
time the Antarctic Bottom Water has moved this far north, it has a
much thinner nepheloid layer associated with it and thus seems to
have already deposited most of the suspended material Ewing found to
the south. Such a finding also implies that relatively little suspended matter is transferred in near-bottom nepheloid layers to the
western basin of the North Atlantic from the south. This would be
consistent with the fairly sharp regional mineralogic boundaries
Biscaye [1965] found in surface sediments from the South Atlantic.
Data of the present study also indicate that the thick (>1000 m)
nepheloid layers in the Hatteras and Sohm Abyssal Plain areas
[*Eittreim et al.*, 1969; *Betzer and Pilson*, 1971] are generated north
of the study area. The suspended particle loads in the thin (<200 m
thick) near-bottom nepheloid layer at 11°N are about 1/6 those found
to the north over the Hatteras and Sohm Abyssal Plains [*P. E.
Biscaye*, personal communication; *S. Eittreim*, personal communication; *Betzer*, unpublished data].

The depth distributions of suspended particulate matter over the
Mid-Atlantic Ridge (stations 10 and 11) show a thin (<100 m) nepheloid layer near the ridge crest (Figure 6). *POC* contributes a
smaller proportion to the suspended matter in this layer than in the
overlying waters; ten samples collected between 1100 and 3400 m gave
an average of 29% *POC*; the two samples collected 30 m above the
ridge gave an average of 10% *POC*. The contribution of suspended
carbonates to the suspended particulate matter also decreases
slightly in the nepheloid layer at the ridge crest. Thirteen samples collected between 1000 and 3400 m gave an average of 6.6% carbonate; the two near-bottom samples gave an average of 4.5%. The
50-60% increase in suspended particulate matter in the nepheloid
layer near the ridge crest is mainly due, then, to an increase in
what might be termed a refractory fraction comprised of clay minerals, oxides and/or hydroxides. Chemical analysis of the carbonate-
free particulate matter for iron and maganese should indicate

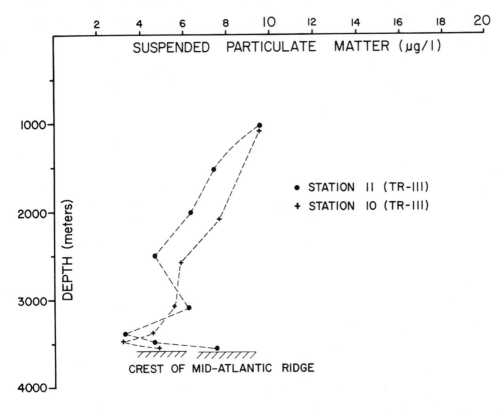

Fig. 6. Depth distributions of suspended particulate matter (μg/ℓ) at two stations over the Mid-Atlantic Ridge at 11°N.

whether the refractory fraction present near the ridge crest is composed of materials which diffused out of or were injected from the ridge into North Atlantic Deep Water [*Bostrom and Peterson*, 1966; and 1969]. In this regard it is interesting to note the increased ratios of β(45)/Total Surface Area (*TSA*) just over the Mid-Atlantic Ridge (Figure 7). This variable is, essentially, light scattering per unit surface area of particulate matter, discussed by *Carder et al.* in this volume. Calculation of the total particle surface area of a sample was carried out using data obtained using a Coulter counter; hence, in this case, all particles <2.3 μm in diameter are not accounted for. In essence, then, a truncated size distribution and a variable β(45)/*TSA* sensitive to the presence of large numbers of finely divided particles are being dealt with. Finely divided iron and manganese-rich particles have been found associated with deposits near the East Pacific Rise [*Bostrom and Peterson*, 1966]. It is possible that an input of such finely divided particles from the ridge to North Atlantic Deep Water is causing

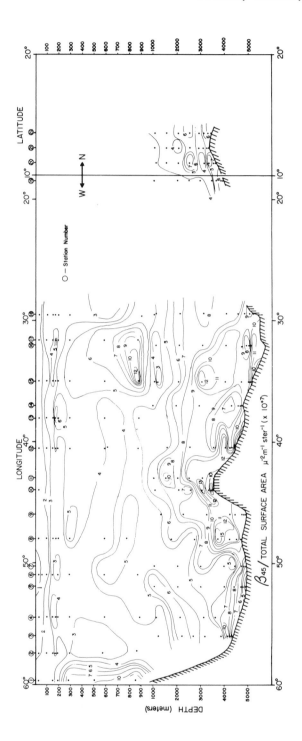

Fig. 7. Contours showing depth distributions of the ratio of light scattering $\beta(45)$ to total particle surface area for a transect of the North Atlantic Ocean at 11°N.

the great distortion (a twofold increase) in $\beta(45)/TSA$ over the Mid-Atlantic Ridge at 11°N. The presence of the high ratios in samples collected from depths as shallow as 1500m at stations 10 and 11 may mean that some of this material was entrained by North Atlantic Deep Water north of the study area, where the ridge is much shallower, and then advected southward. North of the study transect, the axis of the Mid-Atlantic Ridge trends northwestward. In this regard, it is interesting to note the increase in $\beta(45)/TSA$ just west of the ridge. The distortion in this ratio at station 8 may also be due to a southward transport of fine-grained ridge material by North Atlantic Deep Water.

Two additional stations were sampled over the Mid-Atlantic Ridge at 25-26°N on R/V *Trident* cruise TR-114. These stations were near a zone of high heat flow and anomalous manganese-iron crusts [*Scott et al.*, in preparation, 1973]. The near-bottom samples from these two stations also show significant increases in suspended particulate matter. Iron and manganese analyses now in progress will help in determining whether the suspended matter from these two stations, as well as that collected at 11°N, has been affected by the ridge.

Costin [1970] and *Carder and Schlemmer* [1973] have found nepheloid layers in the Gulf of Mexico and Carribean Sea associated with the Subtropic Underwater (75-200 m) and the oxygen minimum zone (400-500 m). The data for suspended particulate matter from cruises TR-111 and TR-127 to the Yucatan Channel are presented in Table 1.

TABLE 1. Comparison of Suspended Particulate Matter Levels ($\mu g/\ell$) in Water Masses of the Western Atlantic Ocean and Northwestern Carribean Sea.

	Western Atlantic (Cruise TR-111)	Yucatan Channel (Cruise TR-127)
Subtropic Underwater	22.8 (n=24)	40 (n=15)
Oxygen Minimum Water	18.5 (n=6)	69 (n=4)
Antarctic Intermediate Water	16.4 (n=6)	29 (n=16)

n = the number of samples taken in each water mass.

In the western Atlantic Ocean adjacent to the Antilles, the suspended loads and $\beta(45)$ are all relatively low in Subtropic Underwater, oxygen minimum water, and Antarctic Intermediate Water. Sampling of the same water masses in the Yucatan Channel indicated a much larger load of particles. In transit through the Carribean Sea, then, a large quantity of particles must be entrained by these three water masses. Three potential sources of the particles are: (1) primary production; (2) rivers emptying into the Carribean and adjacent seas (Magdalena, Orinoco and/or Amazon Rivers); and (3) scouring of the abundant carbonate banks by bottom currents in the northwestern Carribean Sea. Chemical and mineralogical investigations of the suspended particles in the Yucatan Channel should help to identify the more important particle sources for each of these water masses. That is, a large clay fraction (determined by aluminum/suspended particulate matter ratio) would suggest the rivers as a source, whereas, large amounts of calcite (determined by x-ray diffraction analysis, atomic absorption spectrophotometry for Ca/Mg and Ca/Sr ratios in suspended matter after digestion with acetic acid) would suggest that open-ocean producers (coccolithophores) had added large amounts of material. If the carbonate banks were a significant source of particles, a large aragonite content (determined by x-ray diffraction, atomic absorption spectrophotometry for Ca/Mg and Ca/Sr ratios in the suspended matter after digestion with acetic acid) would be evident.

The depth distributions of suspended particulate matter and $\beta(45)$ both show significant increases between 3000 and 4000 m near the South American side of the western basin (stations 3 and 4; Figures 2 and 3). The potential temperature/salinity characteristics of these water samples are consistent with this being a southerly extension of the Western Boundary Undercurrent. Here, the problem is the same as that faced by oceanographers at Cape Hatteras: the classic conservative properties (potential temperature and salinity) are not sufficiently diagnostic for differentiation of the Western Boundary Undercurrent from other North Atlantic Deep Water [*Richardson*, 1971]. In this case, non-conservative tracers may help identify the path taken by the Western Boundary Undercurrent after it moves off in a southeasterly direction from the Cape Hatteras area [*Barrett*, 1965].

The chemical composition of the particulate matter at 11°N may help determine whether the water between 3000 and 4000 m is an extension of the Western Boundary Undercurrent. The chlorite-rich clay mineral assemblages over which the Western Boundary Undercurrent flows [*Biscaye*, 1965], prior to its losing contact with the ocean bottom, may provide the chemical tracers needed. If the water found at 11°N is an extension of the Western Boundary Undercurrent, then some residual chlorite suspended in the water should be evident. This suspended chlorite would have a major affect on the chemical composition of the suspended materials carried by the Western Boundary Undercurrent, providing marked contrast with the Antarctic Bottom Water below it which flows over a chlorite-poor, kaolinite-rich assemblage at 11°N [*Biscaye*, 1965]. It is important to note that the Antarctic Bottom Water and Western Boundary Undercurrent flow over sediment having essentially the same weight percent illite and montmorillonite prior to arriving at 11°N. In addition, atmospheric dust carried from Africa and Europe has minor amounts of chlorite (10%), making it unlikely that this source of chlorite could mask that chlorite carried by the Western Boundary Undercurrent. In Table 2, in a comparison of the chemical composition of kaolinite and chlorite (based on *Degens*, 1965, p. 24), two

fairly powerful chemical ratios appear: iron/suspended particulate
matter and magnesium/suspended particulate matter.   That is, if
chlorite is found between 3000 and 4000 m at stations 3 and 4, these
ratios would be expected to differ from those for samples collected
from lower in the water column.   The chemical analyses which were
carried out on the carbonate fractions provide some **evidence** that
this is the case.   Between 3000 and 4000 m at stations 3 and 4, there
are twofold to threefold increases in the concentrations of particulate magnesium, but not calcium.   Along the remainder of the transect,
the analyses of the acetic acid digestions indicate a parallel relationship of particulate calcium with particulate magnesium and a
calcium/magnesium ratio close to that of low-magnesium calcites,
suggesting that these two elements are derived predominantly from
carbonates.   At stations 3 and 4, the large increases in **magnesium**
without concomitant increases in calcium may indicate that chlorite
was present and that some was dissolved by the acetic acid [*Chester
and Hughes*, 1967].   Definitive proof of the presence of suspended
chlorite must await chemical analyses of the suspended matter, freed
of its carbonate, for magnesium and iron.

Distinct near-bottom nepheloid layers at between 30 and 100 m
thickness are found at stations 19, 20, 21, and 22 on the African

TABLE 2.   The Chemical Composition (% by Weight) of
Kaolinite and Chlorite (based on *Degens*, 1965).

	Kaolinite	Chlorite	Chemical Ratio	$\frac{\text{Chlorite}}{\text{Kaolinite}}$
SiO_2	45	26.6		0.6
Al_2O_3	38	25		0.6
Fe_2O_3	0.8	—	11	
FeO	—	8.7		
MgO	0.08	26.9		336
CaO	0.08	0.28		3.5

continental rise, 48 km from the shelf.   The samples taken 30 m from
the bottom at these stations had an average suspended load of 19
μg/ℓ — over four times the average of 4.4 μg/ℓ found in samples
taken 30 m from the bottom at stations 12, 14, 15, amd 18 in the
central eastern basin.   The distinct difference between waters near
the African continental slope and those of the deep eastern basin is
also reflected in the light-scattering measuremnts β(45) for the
same samples: stations 19-22 gave an average of 4.4 ± 0.4; the
abyssal stations gave an average of 3.6 ± 0.8.   In addition, the
total particle load in water, taken between 1000 m and the bottom,

is 7.9 µg/ℓ in 40 samples from the abyss (stations 12, 14, 15, 16 and 18) compared to 14.5 µg/ℓ in 32 samples near the African continental slope (stations 19, 20, 21 and 22). The difference between these means is significant at the 99% confidence level.

The enrichment of particulate matter below 1000 m depth near the African continental slope might be due to the following: (1) biological debris arising from upwelling along the northwest African coast [*Sverdrup et al.*, 1942]; (2) dust transported from the Sahara Desert by the easterly trade winds (axis at 11°N, the latitude of our stations, in February, at the time of sampling); (3) terrigenous material from the west African shelf. The chemical and optical data gathered at stations 19-22 can help determine which of these sources is probably dominant.

POC and *PC* levels at stations 19 and 20 are very low, averaging 16% and 2.7% of the suspended particulate matter, respectively, with refractory material comprising 81.3%. Thus the significant increases in particulate matter on the continental rise are probably not due to pelagic input of biological material from shallower waters above. Inputs to the ocean from both the atmosphere [*Delany et al.*, 1967] and west African shelf [*McMaster et al.*, in preparation, 1973] are dominated by refractory inorganic matter (clays, feldspars, and quartz). Compositional data from a near-bottom nepheloid layer found on R/V *Trident* cruise 112 over a 450-km stretch of the continental shelf off Liberia and Sierra Leone shows a very low *POC* component (13.5%), a lower *PC* component (7%), and a very large refractory component (79.5% [*McMaster et al.*, in preparation, 1973]) similar to the composition of the suspended material at stations 19 and 20. The suspended terrigenous matter in this nepheloid layer is being carried toward stations 19 and 20 by a bottom current found by *McMaster et al.* [in preparation, 1973] who estimated the annual flux of suspended sediment in this bottom current to be very large, about 0.6% of the sedimentation rate for the entire eastern basin of the North Atlantic Ocean. Thus, both physical and chemical oceanographic evidence support the idea that terrigenous material carried from the west African shelf by this bottom current is a major source of particles for the nepheloid layer near the African continental slope.

The optical and physical data are also consistent with this explanation. The slopes of the particle-size distribution (number versus diameter) plotted on a log-log scale are 3.27 ± 0.19 and 2.99 ± 0.10 for samples taken over the abyss (stations 12, 14, 15, 16 and 18) and African continental slope (stations 19, 20, 21 and 22), respectively. Assuming the difference is significant, the relatively low slope associated with samples taken in deep water near Africa means greater numbers of large particles are present here than in the deep eastern basin. Lower ratios of $\beta(45)/TSA$ are found near the African continental slope than to the west, implying that great numbers of coarser particles are suspended in this area. Both these findings support the hypothesis that significant amounts of shelf-derived materials are contributed to this ocean area since shelf-derived particles are significantly coarser than atmospheric-borne particles [*Delany et al.*, 1967; R. *Duce*, personal communication].

The northwest trend to the contours of carbonate content in surface sediments in this area [*Turekian*, 1968] also supports the idea that large amounts of terrigenous material are being delivered to the deep ocean from the southeast or the direction of the shelf.

Atmospheric fall-out and wash-out do contribute significant amounts of suspended matter to the deep eastern basin of the North Atlantic Ocean [*Delany et al.*, 1967]. It seems likely, however, that with proximity to the African continental slope, material derived from the continental shelf takes on greater importance than the atmospheric component. A more reliable differentiation of the materials from the atmosphere and shelf will be made later by determining the chemical composition of the refractory particulate fraction carried by each of these sources, as well as the refractory fraction collected at stations 19, 20, 21 and 22.

At stations 1, 2, 4, 6, 8, 10, 12, 13, 15 and 18, samples were taken at 30 m depth. Expendable bathythermograph traces from all these stations indicated that the seasonal thermocline was well below this depth — in most cases, close to 95m. All samples, therefore, were taken from the mixed layer. The weight of suspended particulate matter ranged from 5.0-79.6 µg/ℓ with an average of 31 µg/ℓ, 4 times the deep-water average of 8.4 µg/ℓ. This finding agrees with recent data for the open northeastern Atlantic Ocean where the average suspended particle load in the mixed layer was also 4 times that in deep-water [*Copin-Montegut and Copin-Montegut*, 1972].

Systematic increases and decreases in suspended matter are evident in the mixed layer all across the Atlantic (Figure 2). From west to east along the cruise track, particle mass is high, decreases sharply to a low, and then increases to another maximum. It has already been pointed out that, during the winter months, 11°N is the approximate axis for the easterly trade winds which deliver substantial amounts of particulate matter to the ocean [*Arrhenius*, 1963]. The cyclic data pattern and substantial differences in suspended matter concentrations in the mixed layer over the transect suggest that the delivery of atmospheric particulate matter to the ocean is uneven. The available data for the composition of the suspended particulate matter suggest an atmospheric rather than a biological source. That is, the large amounts of suspended matter in the mixed layer at stations 10, 12 and 18 have low *POC* and *PC* concentrations and an average refractory fraction (clay, oxides, hydroxides) of 73%. The other 8 stations all have much higher *POC* contributions relative to the suspended matter (some, but not all, have increased *PC* fractions) and a refractory fraction averaging 27%. Further, the increase in the ratio of $\beta(45)$ to suspended particulate matter in the eastern basin (Figure 4) suggests presence of very fine, optically active particles which were not entirely separated from the water samples by the 0.4µ pore-size filters used. These fine particles are probably more abundant in the eastern basin than in the western basin because it is closer to the source of atmospheric dust. The high concentrations of suspended particulate matter at stations 10, 12 and 18 may have resulted from a wash-out of dust from the Sahara Desert during a rain storm and/or an increased fall-out of dust from the Sahara Desert under the changing axis of the easterlies.

SUMMARY

In summary, the following generalizations can be made.

(1) A combined optical, physical and chemical investigation of suspended particulate matter was carried out on a transect of the North Atlantic Ocean at about 11°N.

(2) The development of a near-bottom nepheloid layer in the western basin of the North Atlantic Ocean is directly related to the flux of Antarctic Bottom Water over the bottom. The lack of such a layer in the eastern basin is most likely due to the slow circulation of its deep water.

(3) Distortions in optical, physical and chemical parameters were found near the South American side of the western basin, and chemical investigations are proposed to determine whether the Western Boundary Undercurrent is the cause of these distortions.

(4) Increases in the proportion of refractory materials were found near the crest of the Mid-Atlantic Ridge, representing, possibly, an input of ridge materials to the deep ocean.

(5) A comparison of three water types was made in both the western Atlantic Ocean and the Yucatan Channel. In each case much greater amounts of suspended particulate matter were present in the Yucatan Channel, showing that a significant build-up of particulate matter takes place in shallow and intermediate water masses passing through the Caribbean Sea. The several possible sources for this particulate matter include: several major South American rivers (Magdalena, Orinoco, Amazon) near the Caribbean Sea; the abundant, shallow carbonate banks in the northwestern Caribbean Sea; and the marine biosphere. The material from these sources has not only distinctly different chemical compositions (silicon-, aluminum- and iron-rich suspended matter from the rivers versus calcium- and magnesium-rich material from the carbonate banks and phytoplankton) but also distinctly different mineralogies (clay-rich suspended matter from the rivers, aragonite-rich suspended matter from the carbonate banks, calcite-rich suspended matter from the phytoplankton) which might be used to differentiate them.

(6) Deep nepheloid layers found near the African continental rise have chemical compositions and particle sizes which relate them to a terrigenous source on the west African shelf.

(7) Suspended matter maxima in the mixed layer over the transect at 11°N are most likely accounted for by input of atmospheric dust.

ACKNOWLEDGEMENTS

The officers and crew of the R/V *Trident*, the marine technicians of the University of Rhode Island and Bob Duce, the chief scientist of *Trident* cruise TR 111, made it possible for us to work efficiently at sea. Without the able assistance of the chief engineer of the *Trident*, John Symonds, much of the work would have been seriously curtailed. Many valuable suggestions for improving the manuscript were made by Susan B. Betzer. The work was supported by the Office of Naval Research under Contract N00014-72-A-0363-0001.

REFERENCES

Arrhenius, G., Pelagic sediments, in *The Sea*, vol. 3, edited by M.N. Hill, pp. 655-727, Wiley Interscience, New York, 1963.

Betzer, P. R. and M. E. Q. Pilson, Particulate iron and the nepheloid layer in the western North Atlantic, Caribbean and Gulf of Mexico, *Deep Sea Res.*, *18*, 753, 1971.

Barret, J. R., Jr. Subsurface currents off Cape Hatteras, *Deep Sea Res.*, *12*, 173, 1965.

Beardsley, G. F., Jr., The polarization of the near asymptotic light field in sea water, Ph.D. thesis, Cambridge, Massachusetts Institute of Technology, 1966.

Biscaye, P. E., Mineralogy and sedimentation of recent deep-sea clay in the Atlantic Ocean and adjacent seas and oceans, *Geol. Soc. Amer. Bull.*, *76*, 803, 1965.

Bostrom, K. and M. N. A. Peterson, Precipitates from hydrothermal exhalations on the East Pacific Rise, *Econ. Geol.*, *61*, 1258, 1966.

Bostrom, K. and M. N. A. Peterson, The origin of aluminum-poor ferromanganoan sediments in areas of high heat flow on the East Pacific Rise, *Mar. Geol.*, *7*, 427, 1969.

Carder, K. L., Particles in the eastern equatorial Pacific Ocean: Their distribution and effect upon optical parameters, Ph.D. thesis, Oregon State University, Corvallis, 1970.

Carder, K. L. and F. C. Schlemmer II, Distribution of particles in the surface waters of the eastern Gulf of Mexico: An indicator of circulation, *J. Geophys. Res.*, *78*, 6286, 1973.

Chester, R. and M. J. Hughes, A chemical technique for the separation of ferro-manganese minerals, carbonate minerals and adsorbed trace elements from pelagic sediments, *Chem. Geol.*, *2*, 249, 1967.

Copin-Montegut, C. and G. Copin-Montegut, Chemical analyses of suspended particulate matter collected in the northeast Atlantic, *Deep Sea Res.*, *19*, 445, 1972.

Costin, J. M., Jr., Visual observations of suspended-particle distribution at three sites in the Caribbean Sea, *J. Geophys. Res.*, *75*, 4144, 1970.

Degens, E. T., *Geochemistry of Sediments*, Prentice-Hall, Inc., Englewood Cliffs, N. J., p. 24, 1965.

Delany, A. C., A. C. Delany, D. W. Parkin, J. J. Griffin, E. D. Goldberg and B. E. F. Reimann, Airborne dust collected at Barbados, *Geochim. et Cosmochim. Acta.*, *31*, 885, 1967.

Eittreim, S. L., Suspended sediment in the Northwest Atlantic Ocean, Ph.D. thesis, Columbia University, New York, 1970.

Eittreim, S. L., P. M. Bruchhausen and M. Ewing, Vertical distribution of turbidity in the south Indian and south Australian Basins, in *Antarctic Oceanology II*; The Australian-New Zealand Sector, Antarctic Res. Ser., vol. 19, edited by D. E. Hayes, pp. 51-58, AGU, Washington, D.C., 1972.

Eittreim, S. L., M. Ewing and E. M. Thorndike, Suspended matter along the continental margin of the North American Basin, *Deep Sea Res.*, *16*, 613, 1969.

Ewing, M. and S. Connary, Nepheloid layer in the North Pacific, *Geol. Soc. Amer. Mem.* *126*, edited by J. Hays, pp. 41-82, 1970.

Ewing, M., J. Ewing and X. LePichon, Sediment transport and distribution in the Argentine Basin, 4, nepheloid layer and bottom water circulation (abstract), *Trans. Am. Geophys. Un., 48,* 141, 1967.

Fournier, R. O., Observations of particulate organic carbon in the Mediterranean Sea and their relevance to the deep-living coccolithorphorid *cyclococcolithus fragilis, Limnol. Oceanogr., 13,* 693, 1968.

Hunkins, K., E. M. Thorndike and G. Mathieu, Nepheloid layers in the Arctic Ocean, *J. Geophy. Res., 74,* 6995-7008, 1969.

McMaster, R. L., P. R. Betzer, K. L. Carder, L. Miller, and D. W. Eggimann, Suspended particle mineralogy and water masses of the west African shelf adjacent to Sierra Leone and Liberia, in preparation, 1973.

Manheim, F. T., R. H. Meade and G. C. Bond, Suspended matter in surface waters of the Atlantic margin from Cape Cod to the Florida Keys, *Science, 167,* 371, 1970.

Menzel, D. W. and R. F. Vaccaro, The measurement of dissolved organic and particulate carbon in sea water, *Limnol. Oceanogr., 9,* 138, 1964.

Metcalf, W. G., Dissolved silicate in the deep North Atlantic, *Deep Sea Res., 16* (supplement), 139, 1969.

Pak, H., The Columbia River as a source of marine light-scattering particles, Ph.D. thesis, Oregon State University, Corvallis, Oregon, 1970.

Richardson, P. L., Transport and velocity structure of the Gulf Stream at Cape Hatteras, in *The Ocean World,* edited by Michitake Uda, Japan Society for the Promotion of Science, Tokyo, pp. 383-384, 1971.

Scott, M., R. Scott, P. Rona, L. Butler, and A. Nalwalk, Hydrothermal manganese from the Mid-Atlantic Ridge, in preparation, 1973.

Sheldon, R. W. and T. R. Parsons, *A Practical Manual on the Use of the Coulter Counter in Marine Science,* Coulter Electronics, Toronto, 66 pp., 1967.

Spilhaus, A. F., Jr., Observations of light scattering in sea water, Ph.D. thesis, Massachusetts Institute of Technology, Cambridge, Mass., 1965.

Sverdrup, H. U., M. W. Johnson, and R. H. Fleming, *The Oceans,* Prentice-Hall, Inc., Englewood Cliffs, N. J., 1087 pp., 1942.

Turekian, K. K., *Oceans,* Prentice-Hall, Englewood Cliffs, New Jersey, 120 pp., 1968.

Worthington, L. V. and W. G. Metcalf, The relationship between potential temperature and salinity in deep Atlantic water, *Rapp., Cons. Explor. Mer, 149,* 122, 1961.

Wright, W. R., Northward transport of Antarctic Bottom Water in the Western Atlantic Ocean, *Deep Sea Res., 17,* 367, 1970.

LIST OF CONTRIBUTORS AND PARTICIPANTS

NEIL ANDERSON, Office of Naval Research, U. S. Department of the
Navy, Arlington, Virginia 22217

ROSWELL W. AUSTIN, Visibility Laboratory, Scripps Institution of
Oceanography, University of California, San Diego, La Jolla,
California 92037

EDWARD T. BAKER, Department of Oceanography, University of Washington,
Seattle, Washington 98195

N. JAY BASSIN, Department of Oceanography, Texas A & M University,
College Station, Texas 77843

PETER R. BETZER, Department of Marine Sciences, University of South
Florida, St. Petersburg, Florida 33701

PIERRE E. BISCAYE, Lamont-Doherty Geological Observatory, Columbia
University, Palisades, New York 10964

KENDALL R. CARDER, Department of Marine Sciences, University of South
Florida, St. Petersburg, Florida 33701

MICHAEL F. DACEY, Department of Geography and Department of Geological
Sciences, Northwestern University, Evanston, Illinois 60201

DAVID E. DRAKE, Marine Geology and Geophysics Laboratory, National
Oceanic and Atmospheric Administration, Atlantic Oceanographic
and Meteorological Laboratories, 15 Rickenbacker Causeway,
Miami, Florida 33149

DONALD W. EGGIMANN, Department of Marine Sciences, University of
South Florida, St. Petersburg, Florida 33701

STEPHEN L. EITTREIM, Lamont-Doherty Geological Observatory, Columbia
University, Palisades, New York 10964

MAURICE EWING, Earth and Planetary Sciences Division, Marine
Biomedical Institute, University of Texas, Galveston, Texas
77550 (Deceased May 5, 1974)

RICHARD A. FEELY, Department of Oceanography, Texas A & M University,
College Station, Texas 77843

EDWARD S. FRY, Physics Department, Texas A & M University, College
Station, Texas 77843

RONALD J. GIBBS, College of Marine Studies, University of Delaware,
Lewes, Delaware 19958

HOWARD R. GORDON, Physics Department, University of Miami, Coral
Gables, Florida 33124

DONALD HEINRICHS, Office of Naval Research, U. S. Department of the
Navy, Arlington, Virginia 22217

DAVID HODDER, Space Division, North American Rockwell, 12214
Lakewood Blvd., Downey, California 90241

ANTON L. INDERBITZEN, College of Marine Studies, University of
    Delaware, Lewes, Delaware 19958
GUNNAR KULLENBERG, Institute of Physical Oceanography, Copenhagen
    University, Copenhagen, Denmark
DEVENDRA LAL, Physical Research Laboratory, Navrangpura, Ahmedabad
    380009, India; Scripps Institution of Oceanography, La Jolla,
    California 92037
ABRAHAM LERMAN, Department of Geological Sciences, Northwestern
    University, Evanston, Illinois 60201
ALEXANDER MALAHOFF, Office of Naval Reasearch, U. S. Department of
    the Navy, Arlington, Virginia 22217
DEAN A. McMANUS, Department of Oceanography, University of
    Washington, Seattle, Washington 98195
HASONG PAK, School of Oceanography, Oregon State University,
    Corvallis, Oregon 97331
THEODORE J. PETZOLD, Visibility Laboratory, Scripps Institution of
    Oceanography, University of California, San Diego, La Jolla,
    California 92037
WILLIAM M. SACKETT, Department of Oceanography, Texas A & M
    University, College Station, Texas 77843
JERRY R, SCHUBEL, Chesapeake Bay Institute, Johns Hopkins University,
    Baltimore, Maryland 21218
RAYMOND C. SMITH, Visibility Laboratory, Scripps Institution of
    Oceanography, University of California, San Diego, La Jolla,
    California 92037
DEREK W, SPENCER, Department of Chemistry, Woods Hole Oceanographic
    Institute, Woods Hole, Massachusetts 02543
RICHARD W. STERNBERG, Department of Oceanography, University of
    Washington, Seattle, Washington 98195
ROBERT E. STEVENSON, Office of Naval Research, U. S. Department of
    the Navy, La Jolla, California 92037
LAWRENCE SULLIVAN, Lamont-Doherty Geological Observatory, Columbia
    University, Palisades, New York 10964
STEVENS P. TUCKER, Oceanography Department, U. S. Naval Post-Graduate
    School, Monterey, California 93940
JOHN E. TYLER, Visibility Laboratory, Scripps Institution of
    Oceanography, University of California, San Diego, La Jolla,
    California 92037
J. RONALD V. ZANEVELD, School of Oceanography, Oregon State University,
    Corvallis, Oregon 97331

INDEX